Planet Earth

CONTINENTS IN COLLISION

TIME
LIFE
BOOKS

This volume is one of a series that examines the
workings of the planet earth, from the geological
wonders of its continents to the marvels of its
atmosphere and its ocean depths.

Cover
Tinted by an amber sunset, a lava field on
the Canary Islands evidences its molten origins.
The lava oozed from one of the deep magma
sources known as hot spots, which play a part in
the processes that eternally reshape the earth.

CONTINENTS IN COLLISION

By Russell Miller
and The Editors of Time-Life Books

Time-Life Books, Amsterdam

PLANET EARTH

EDITOR: Thomas A. Lewis
Designer: Donald Komai
Chief Researcher: Pat S. Good

Editorial Staff for *Continents in Collision*
Associate Editors: Russell B. Adams Jr. (text);
Peggy Sawyer Seagrave (pictures)
Text Editor: Anne Horan
Staff Writers: Tim Appenzeller, Adrienne George,
Stephen G. Hyslop, John Newton
Researchers: Kristin Baker, Megan Barnett and Feroline
Burrage (principals), Elizabeth B. Friedberg, Robert
Kachur, Stephanie Lewis
Assistant Designer: Susan K. White
Copy Coordinators: Victoria Lee, Bobbie C. Paradise
Picture Coordinator: Donna Quaresima
Editorial Assistants: Caroline A. Boubin,
Annette T. Wilkerson

Correspondents: Elisabeth Kraemer (Bonn); Margot
Hapgood, Dorothy Bacon (London); Miriam Hsia,
Lucy T. Voulgaris (New York); Maria Vincenza
Aloisi, Josephine du Brusle (Paris); Ann Natanson
(Rome). Valuable assistance was also provided by:
Helga Kohl (Bonn); Robert W. Bone (Honolulu);
Cheryl Crooks (Los Angeles); Judy Aspinall,
Millicent Trowbridge (London); Christina Lieberman
(New York); Bogi Agustsson (Reykjavik); Mimi
Murphy, Ann Wise (Rome); Kasuko Yamazaki (Tokyo).

THE AUTHOR

Russell Miller is a British journalist and free-lance writer who frequently contributes to *The Sunday Times* of London. He is the author of three previous Time-Life books: *The Resistance* and *The Commandos* for the World War II series and *The East Indiamen* for The Seafarers series.

THE CONSULTANT

Tanya Atwater, Professor of Geophysics at the University of California at Santa Barbara, has written numerous articles and a textbook on plate tectonics. She is a fellow of the American Geophysical Union and the Geological Society of America, and a member of the Public Advisory Committee on the Law of the Sea of the United States Department of State. She has made many dives in the deep-sea research submersible *Alvin*, including one on which she served as chief scientist.

ISBN 7054 0746 2

TIME-LIFE is a trademark of Time Incorporated U.S.A.

CONTENTS

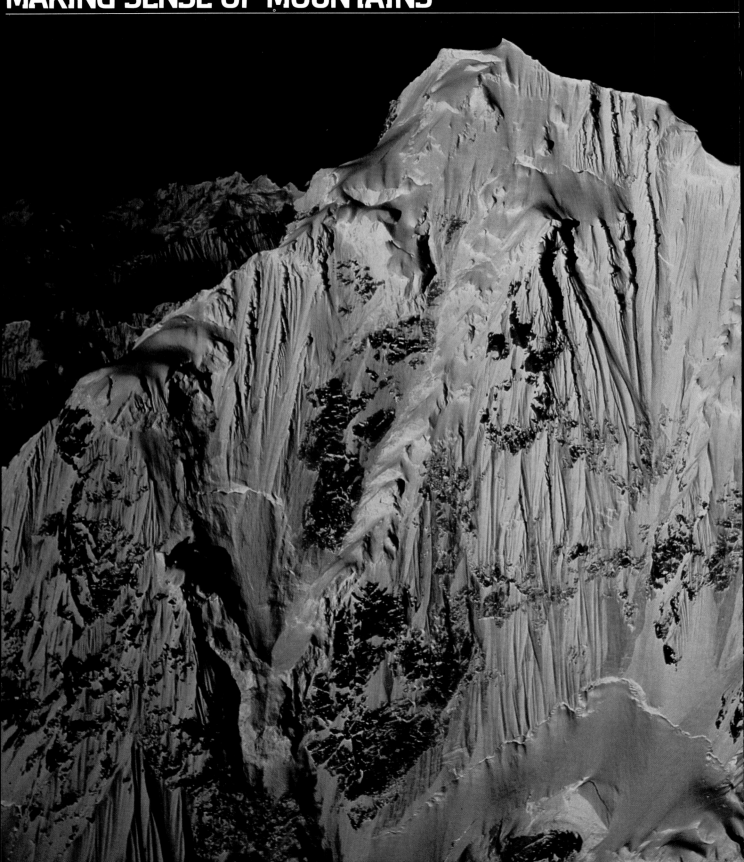

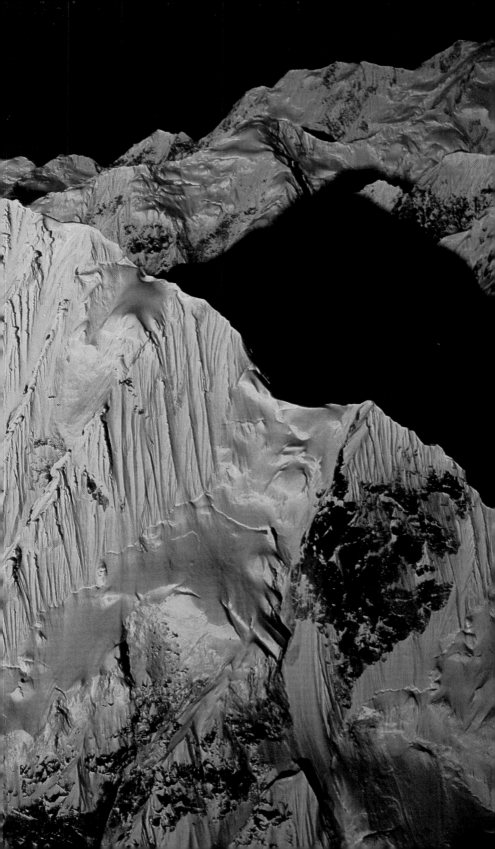

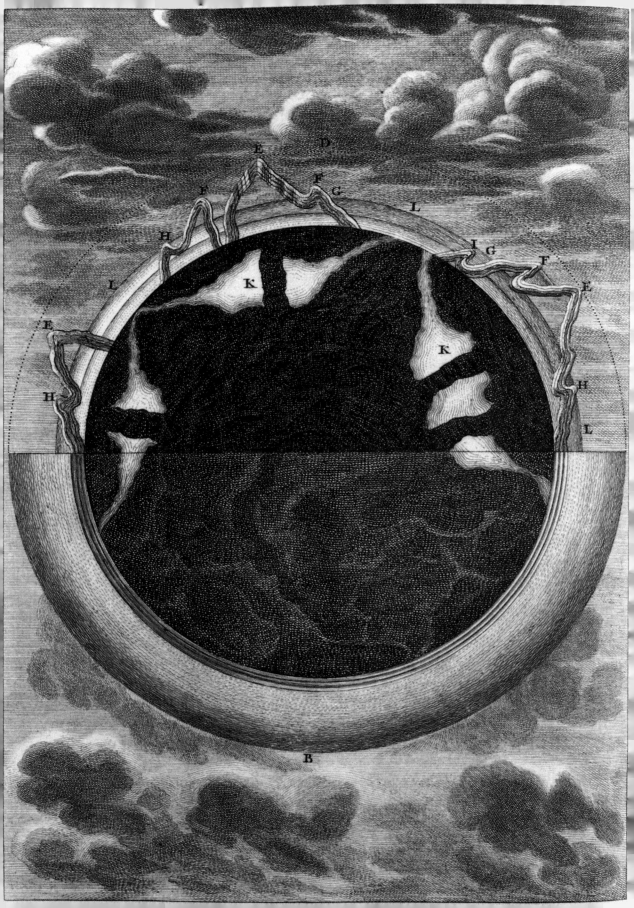

Chapter 1

A MYSTERIOUS AND MOBILE EARTH

Few scholars of the 17th Century enjoyed or deserved the international respect accorded James Ussher, Anglican Archbishop of Armagh, Primate of All Ireland, member of England's Privy Council and confidant of his King, Charles I. Thus it is hardly surprising that the publication of the major work of his life—a rigorous, exhaustive research project that led him to dramatic conclusions about the nature and the history of the earth it-self—had an immediate, profound and lasting influence.

Ussher constructed a chronology for this history, using as a source the Holy Bible, by his lights the only acceptable text on the origin of the planet. (Like other Western scholars, he regarded the ancient Greek, Egyptian and Chinese accounts of the earth's history to be unreliable compendiums of mythology and legend.) The Bible, however, was a story told to other peoples in another time and in different voices and languages. Ussher set himself to sorting out the inconsistencies and contradictions in its narrative to determine the exact chronology of the events detailed therein. The conclusion of his long and difficult work was the publication in 1650 and 1654 of two volumes of detailed analysis.

Heaven and earth had been created, Ussher concluded, "upon the entrance of the night preceding" Sunday, October 23, 4004 B.C. On the following Tuesday the waters had been gathered together and dry land emerged; man and the other forms of life had appeared on the following Friday. The Deluge followed 1,655 years later, with Noah entering the Ark on Sunday, December 7, 2349 B.C., and leaving it on Wednesday, May 6, the following year. Ussher's treatise, *Annales Veteris et Novi Testamenti,* attracted immediate attention and earned widespread acceptance among scientific as well as religious scholars. His chronology soon began to appear in the margins of the Bible itself, as indeed it does today in various concordances and reference editions.

Ussher had with his scholarship confirmed the conventional wisdom of his age: that the earth was at the most a few thousand years old; that its features had been put in place at a time certain; and that they had remained immutable ever since, untouched by any significant force except, possibly, the Deluge. It was a comforting vision of a stable, ordered planet, and Ussher was honored for it by a grateful country, despite the intervening fall of King Charles I. When Ussher died in 1656, England's Lord Protector Oliver Cromwell ordered a state funeral and burial in Westminster Abbey.

Ussher's scheme was subjected to some nagging challenges from the outset, however—as, for instance, when visitors to far-off China brought back

An 18th Century Swiss engraving illustrates the Biblical account of the third day of Creation. As the seas are divided from the dry land, the earth changes from a featureless, water-covered globe (*lower hemisphere*) to a complex planet of mountains, islands and underground caverns.

word that the recorded genealogies of ancient Chinese families extended back not only to before the time of the Deluge, but to before the accepted date of Creation itself. With the onset of the Enlightenment—that great upswelling of rational thought that swept the Western world in the 18th Century—such questions would arise with ever-increasing frequency and would prove ever more difficult to ignore. As scientific observations of the earth accumulated, they began to indicate that the planet was far older than had been imagined and that it had undergone massive alterations since its first days. Through the restless speculations of the 19th Century and the rise of technology in the 20th, the questions and the new answers they spawned would proliferate until, in the 1960s, a new vision of the earth would finally emerge. When it did, it would transform human thought to a degree unprecedented since Charles Darwin's theory of evolution had challenged the accepted notions about the origins of life, or since Niels Bohr's theory of the atom had utterly revised man's perceptions of matter.

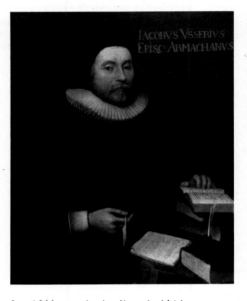

In a 1644 portrait, Anglican Archbishop James Ussher exudes the confidence of his conclusions about the history of the earth. After a painstaking study of the Bible, he announced that the world was created on the night of October 22, 4004 B.C.

Scholars and sages had been aware for thousands of years that the world had not always been as they saw it. In the Fourth Century B.C., Aristotle had puzzled over the existence of fossil marine creatures in rocks high above sea level. Leonardo da Vinci in the 15th Century had recorded the comment: "Above the plains of Italy where flocks of birds are flying today, fishes were once moving in large shoals." Two hundred years later, the brilliant London physicist Robert Hooke concluded from his study of fossils that Britain once had "fishes swimming over it." For people whose scholarship revealed nothing to suggest that the surface of the earth had ever undergone a significant change, the presence of these high-and-dry sea creatures was a perplexing but unavoidable enigma. Hooke speculated that "some Deluge, Inundation, Earthquake or some such other means" had lifted them above sea level, but could offer little to support the notion. Considerations of this abiding mystery were in large part responsible for the rise of the science of geology.

After the great voyages of discovery in the 16th Century, when the first crude maps of the world were drawn, geographers could not help noticing a global puzzle: the strange similarities in the coasts of Africa and South America. In 1620, the English philosopher Sir Francis Bacon commented on the likenesses in his celebrated work *Novum Organum,* calling them "no mere accidental occurrence." A few years later the French moralist François Placet suggested that the New and the Old Worlds had once been joined but were separated by the Flood described in the Bible.

The Deluge was a popular and convenient explanation for the observed changes in the shapes and locations of oceans. But during the 16th and 17th Centuries, scientists became aware of other changes—in landforms and life processes—which were increasingly difficult to reconcile with the church's interpretation of the Biblical account of the Creation. Around 1570, the French naturalist Bernard Palissy had observed that rain, wind and waves were wearing away the continents—so rapidly, he thought, that soon no land would remain unless new rock grew to take its place. Outraged church authorities thought that his remarks questioned the supreme design of the Creator, and denounced him for heresy.

Whether out of piety or fear of notoriety, most early investigators made earnest attempts to reconcile their findings with accepted dogma. The

Although he trained as a physician and died a prelate in 1686, Nicolaus Steno is best remembered as a pioneer of geology. He was the first to understand that sediments are deposited underwater in horizontal layers, with the oldest at the bottom.

Diagrams published by Steno in 1669 explain, in reverse chronological order, his concept of the origin of the earth's features. He thought that level primeval strata *(lower right)* were undermined by water and collapsed *(top right)*, opening chasms in which sediments collected *(lower left)*. A second episode of undermining and collapse, Steno believed, created the cliffs, hills and valleys of the contemporary landscape.

French philosopher and scientist René Descartes concluded in 1634 that the creation of the earth had been the result of ongoing natural processes, but for fear of offending the church, he did not publish his opinions. He was undoubtedly influenced by the fact that his friend Galileo, the brilliant astronomer, mathematician and physicist, had been threatened with torture, had seen his books burned and was living under house arrest in Florence—all for putting forward the heretical view that the planets revolved around the sun.

Nicolaus Steno, a Dane working in Florence in the mid-17th Century as the physician to Grand Duke Ferdinand II, was a gifted amateur geologist, whose studies of the hillsides of Tuscany had convinced him that many rock formations originated as layers of sediment laid down in orderly succession. Thus, he realized, the earth's crust contained a chronological

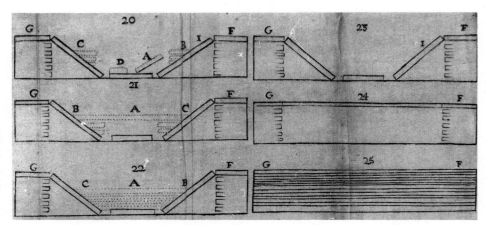

history of geological events that could be unraveled by careful study of each succeeding layer.

Steno was on risky ground here, and in a dissertation published in 1669 he sought to explain his findings strictly in accordance with church dogma. It was a scenario that even Archbishop Ussher, had he still been alive, could not have disputed. The first lands raised by the Creator from the oceans covering the world, Steno claimed, were undermined by gigantic subterranean caverns; with the Deluge, the land collapsed back into the sea, to be replaced by new continents. The post-Flood earth was similarly undermined, but this time the caverns collapsed intermittently to create the existing oceans. The reconciliation of observed evidence with theology could not have been easy for Steno; a few years after the publication of his work, he resolved the conflict by abandoning science and taking holy orders, pursuing both his duties and his vows of poverty so avidly that he died in 1686 of chronic malnutrition.

By that time not only scientists, but members of the clergy as well, were increasingly feeling a need to explain the observed changes in the planet without damaging the doctrine of a single, finite act of genesis. One way was to interpret Noah's flood as a kind of adjustment to the Creation, and in a 1681 book titled *The Sacred Theory of the Earth,* the English cleric Thomas Burnet offered a cataclysmic account of the Deluge with geological overtones.

According to Burnet, the primeval earth was covered by a smooth crust of land over a layer of water. The heat of the sun, he said, had opened deep

Embedded in marble, the coiled shells of mollusk fossils called ammonites form an ancient mosaic. The presence of such relics of marine life in the rocks of high mountain ranges posed a daunting riddle for early students of the earth's history.

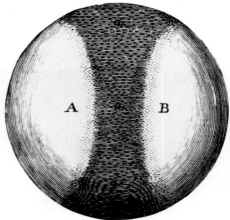
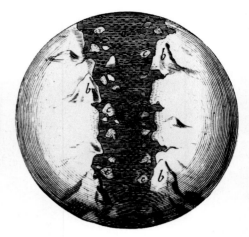

cracks in the crust, through which the underlying water erupted when God's wrath descended on the earth. Huge blocks of land crashed into the depths, some of them upending to form mountains and continents, some of them sinking completely to form the floors of the oceans. The remaining landscape was a great ruin, Burnet concluded, that had undergone no significant change since.

Despite the prodigious mental agility of men like Burnet and Steno, the longer scientists looked at the world around them and the more evidence they found of ongoing processes, the harder it became to pay even lip service to the notion of a static, unchanging planet. Finally, in the mid-18th Century, one of France's leading naturalists boldly turned his back on church doctrine and postulated a new, cosmic vision. Georges LeClerc, the Comte de Buffon, suggested that the earth began as a white-hot mass torn from the sun by some astronomical force; the moon, he said, was in turn ripped from the earth. As the earth cooled, he believed, the oceans drained into deep submarine fissures, slowly exposing the continents. The fact that the continents were formerly the bottom of the sea was incontestable, he wrote, for the "spoils of the ocean are found in every place."

LeClerc's theories so outraged the theological faculty at the Sorbonne that they wrested from him an agreement to withhold publication and to abandon his heretical ideas. Thirty years later he reneged on this agreement and outlined his ideas in a book titled *Epochs of Nature.* His conclusions about the formation of the earth and the solar system were soon discredited but, by publishing them, he had opened new territory to unfettered logic and investigation. And in an aside that was to prove more durable than his basic tenets, LeClerc suggested that something—perhaps the fabled lost continent of Atlantis—might once have joined Ireland and America, for "in Ireland there are the same fossils, the same shells, as appear in America and some of them are found in no other part of Europe."

In 1782 Benjamin Franklin, having helped to frame the Declaration of Independence and the United States Constitution, expressed his opinions on the nature of the earth. He had seen "oyster shells mixed in the stone" of a mountain in Derbyshire, England, and after deliberating on how they might have been raised from their former level below the waves, he arrived at a conclusion that was typically original and durable. The crust of the earth, he said, must be a shell floating on a fluid interior; "thus the surface of the globe would be capable of being broken and disordered by the violent movements of the fluids on which it rested." The astonishing idea, like so much of Franklin's thinking, would be appreciated only by later generations.

Toward the close of the 18th Century, opponents of the Biblical "catas-

Illustrations from a 1684 edition of Thomas Burnet's *Sacred Theory of the Earth* link the formation of the earth's topography to the Biblical Deluge. In Burnet's scenario, a portion of the earth's surface fractured *(left)* and underground waters spilled out to flood the entire globe. Then the water drained back into the collapsed area to form oceans *(center),* which eventually receded enough to reveal the mountainous, island-dotted coastlines of the continents *(above).*

trophe" doctrine found a new ally in James Hutton, a doctor, farmer and naturalist who would later be described as the "founder of modern geology." Hutton set out to ascertain the length of time the earth had existed as a "habitable world," the physical changes it had undergone and what he might predict for the future. From observations of rock formations in his native Scotland, he deduced that the earth continues to be formed and re-formed by natural processes—accumulation, erosion, heating and folding—that have been operating in the same general manner throughout many geological ages. From the barley fields of his farm in Berwickshire, Hutton had watched streams scouring the hills, carrying sediment down to the sea, and had reflected that "old continents are wearing away and new continents are forming in the bottom of the sea."

The earth's history and future, Hutton declared, could be understood not by a literal interpretation of Genesis, but by probing natural processes, the "actual causes," of geological phenomena. "No powers are to be employed," he wrote, "that are not natural to the globe, no action to be admitted except those of which we know the principle, and no extraordinary events to be alleged in order to explain a common appearance."

One of the most remarkable things about Hutton's theory was its perception of the vastness of geological time. To nature, he wrote, time is "endless and as nothing." When scientists seek to limn the long history of the planet, he observed, "we come to a period in which we cannot see any farther." But this did not mean, according to Hutton, that they had thus marked the beginning of time; on the contrary, they had merely reached the limit of human comprehension. A possible end to the world was likewise beyond the perception of man. Wrote Hutton: "We find no vestige of a beginning—no prospect of an end." This—at a time when many still clung to the chronologies of Bishop Ussher—was a truly breathtaking pronouncement.

Hutton's ideas first appeared in relatively obscure publications; then in 1805 his biographer, John Playfair, published *Illustrations of the Huttonian Theory of the Earth,* a book that greatly simplified Hutton's rather ponderous style and made his theories more accessible. As a result, Hutton's cogent rejection of the notion that geological phenomena must be explained strictly in accordance with the Biblical account of Creation began to impress more and more readers. The pioneering Scottish geologist thus had a liberating influence on the fledgling earth sciences, but they were indeed fledgling; the debate between the dogmatists and the investigators had dominated three centuries of exploration and research, during which the overall understanding of the planet had been advanced hardly at all. Even the fundamental question that had perplexed Aristotle—why marine fossils are found on land—remained unanswered. And as they exercised their new freedom and pursued their investigations, researchers like the German explorer and naturalist Alexander von Humboldt succeeded mainly in raising new questions of an ever wider scope.

One of the last great universal scientists, Humboldt studied the geographical distribution of plants and the causes of meteor showers with equal verve. He also became a popular hero through his exploits as an explorer. His most fruitful trip began in 1799, when he and the French botanist Aimé Bonpland embarked for South America. Landing in what is now Venezuela, Humboldt and his companion spent five years traveling and amass-

In a contemporary caricature, 18th Century Scottish geologist James Hutton peers at a cliff in search of evidence to support his controversial ideas about geological processes. The visages scowling at Hutton from the rocks may represent his theory's many opponents.

ing information on landforms, plant and animal life, and weather in what has been called "the scientific discovery of America." Humboldt had an eye for global unities, and during this expedition he realized that the similarities between the Atlantic coastlines of Africa and South America extended far beyond their apparent "fit": Several mountain ranges running parallel with the Equator seemed to end at South America's eastern coast, then resume on the western coast of Africa, and there were striking resemblances in the geological strata of the two continents. The mountains of Brazil, for example, were the same as those of the Congo; the huge Amazon plain was directly opposite the geologically identical plains of upper Guinea in Africa. Furthermore, Humboldt noted, some of the mountain chains of North America seemed to have geological similarities to mountain chains in Europe. Here at last was clear evidence that, as Sir Francis Bacon had suggested nearly 200 years earlier, the likeness between the continents was "no mere accidental occurrence." Humboldt pleaded with fellow scientists for more research into the significance of these global patterns, complaining that "we examine the stones but not the mountains; we have the materials but ignore how they fit together with each other."

Unfortunately, having gathered the evidence, Humboldt then drew the wrong conclusion: He decided that the Atlantic Ocean was "only a valley,"

Botanical specimens, notes, sextants and magnifiers crowd a rude table at the jungle camp of Alexander von Humboldt *(seated)* and Aimé Bonpland in an idealized scene from their 1799-1804 scientific expedition to South America. Geologic observations made during the trip convinced Humboldt that a land bridge once connected South America and Africa.

excavated by erosion and filled with the water of Noah's flood. His theory was soon largely discredited in scientific circles by the wider circulation of James Hutton's arguments that the earth's topography was not the result of ancient catastrophes, but of the same geological processes that were evident to modern man.

Returned from his travels, Humboldt spent the last 25 years of his life writing *Kosmos,* one of the most ambitious scientific works ever published. In five volumes, Humboldt attempted to provide a complete picture of the material world, enriched with every fact then known to earth scientists. The work culminated the lifelong quest for unity that had led to Humboldt's far-reaching geological speculations.

Soon after Humboldt presented his theory about the origin of the Atlantic Ocean, the eminent French biologist Jean-Baptiste Lamarck came up with a novel theory of earth history that explained why so many marine fossils were found on dry land. Lamarck suggested a kind of continental migration around the globe during almost "inconceivable" time spans, in which the land masses were scoured away by tides along their eastern coastlines while sediment deposits extended them at an equal rate along their western shores. Thus, while the continents themselves had not actually moved, the oceans had described a number of complete circumferences of the globe and had previously covered all the continents not once, but several times.

Like Hutton, Lamarck had considerable appreciation for the immensity of time. As Lamarck put it: "Time is insignificant and never a difficulty for Nature. It is always at her disposal and represents an unlimited power with which she accomplishes her greatest and smallest tasks." Unsupported by data, Lamarck's theory elicited no enthusiasm from his fellow scientists, and he was unable to find a publisher for the book, *Hydrogéologie,* in which he expounded it. Determined to see it dignified by print, he paid to have 1,025 copies published privately in 1802, but to his great mortification, he sold very few. Lamarck died blind and poor, in Paris, in 1829 and his books were later sold off at public auction.

An English lawyer, Sir Charles Lyell, proved to be considerably more successful than Lamarck at publishing opinions on geology. Born on his family's Scottish estate in November 1797, Lyell developed an early interest in the natural sciences; when still a schoolboy, he had amassed an impressive collection of butterflies and aquatic insects. Later, he attended Oxford University, where he dutifully studied the classics—and also steeped himself in geological theory. After graduating from Oxford, Lyell moved to London and took up the study of law, but his weakening eyesight made reading difficult and he began to spend more and more of his time doing geological field work. In 1825—the year he was admitted to the bar—he published his first scientific paper, a study of fresh-water limestone formations. More papers followed, and by 1827 Lyell had put aside the practice of law in favor of geology.

Like Hutton, Lyell believed that every feature of the earth's topography was the result of natural processes that were still at work, and he traveled widely in search of supporting evidence. He was particularly intrigued by his studies of Mount Etna and became convinced that this massive volcano had been formed not by some singular cataclysm but by a long succession of eruptions that continued in his own day. In 1830, he presented the argu-

Amidst astrological symbols and luxuriant foliage, Genius unveils Nature on this 1847 medal commemorating the publication of Humboldt's five-volume scientific work, *Kosmos.*

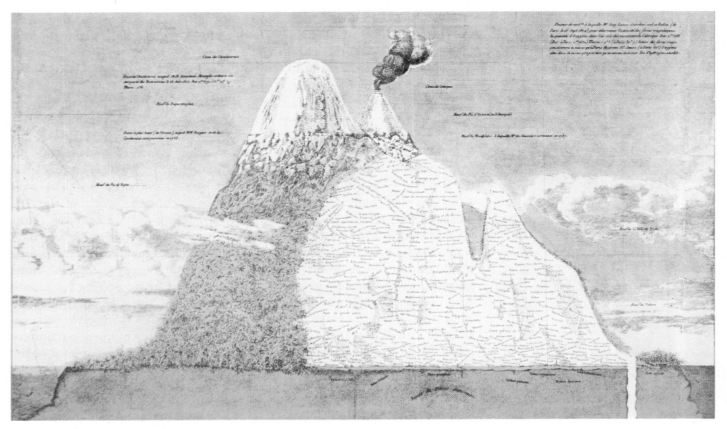

A profile of the Ecuadorian volcano Mount Chimborazo, plotted by the German geographer Alexander von Humboldt, details the plant life that flourished at various altitudes. As Humboldt ascended the mountain in 1802, he noted that the tropical vegetation gradually gave way to arctic species near the summit.

ment in the first volume of his three-volume *Principles of Geology*, a land-mark work that was widely read by scientists. In it, Lyell demonstrated how such natural forces as erosion, sedimentation, earthquakes and volcanic eruptions had altered—and continued to alter—the surface of the earth. He revised and updated his books several times, and in the 1840s and 1850s he traveled frequently to the United States, where his writings and lectures strongly influenced a generation of American geologists.

Unlike Lamarck—whose work he had read—Lyell did not believe that the continents were capable of lateral movement. And while this notion continued to be raised from time to time, it was invariably ridiculed by the scientific establishment, for by the mid-19th Century a new geological theory, which answered at least some of the persistent questions, had gained broad and enthusiastic acceptance. James Dwight Dana, a respected professor of geology at Yale University, declared that the earth was cooling and contracting from a previous molten state and that all geological features— the great mountain ranges included—were caused by its contraction. This was to become the dominant geological theory of the era, for it offered a plausible explanation for the twisted and tortuous rock formations found all around the world.

Dana viewed the continents and deep oceans as permanent features of the earth's crust, fixed at the beginning of geological time. The continents were areas that had cooled first; subsequent contraction had lowered the ocean beds. As the interior of the earth had shrunk, enormous pressures had been exerted on the edges of the continents, precipitating the upheaval of great mountain ranges—the Appalachians, Rockies and Andes—rather like wrinkles appearing on the skin of a drying apple. Enthusiasts of the new theory calculated that, in cooling to its present temperature from a molten state during a period of about 100 million years, the earth's circumference had contracted by hundreds of miles.

According to Dana, the marine fossils that were found on land were the remains of shallow-water creatures whose presence could be explained by minor fluctuations in sea level. But some fundamental questions remained unanswered—such as why mountains had been formed in a few long, widely separated belts, instead of everywhere on the surface like the wrinkles on the apple, and why mountain building had occurred in short, violent epochs separated by long periods of inactivity. Even so, the idea of a cooling and contracting earth satisfied most geologists. "What we are witnessing," wrote the leading Austrian geologist Eduard Suess, "is the collapse of the world."

Suess, professor of geology at the University of Vienna, was a disciple of Dana's contracting-earth theory. In his masterly four-volume work, *The Face of the Earth*, published between 1885 and 1909, he described a cooling earth that was made up of three layers: a central core, an intermediate mantle and a topmost crust. Suess believed that the crust had originally enclosed the entire surface of the earth, but that as the interior shrank, great crustal slabs collapsed into the mantle and created ocean basins. "The crust of the earth gives way and falls in; the sea follows it," he wrote. "It is in the history of the seas that we discover the history of the continents."

The new theory offered a way to explain many of the puzzling biological findings of the 19th Century. Patient investigation of the worldwide distribution of plants and animals had found identical species of snakes, turtles and lizards, none of which could possibly have crossed an ocean, in both Africa and South America. Fossils of the *Glossopteris* fern were uncovered in India and Australia, and fossil remains of *Mesosaurus*, a small reptile that lived 270 million years ago, were found in Brazil and South

Austrian geologist Eduard Suess colored this map in 1909 to show how he thought ancient supercontinents had been grouped. Suess, whose fiery political career inspired the caricature above, believed that land links between the continents later subsided into the ocean basins.

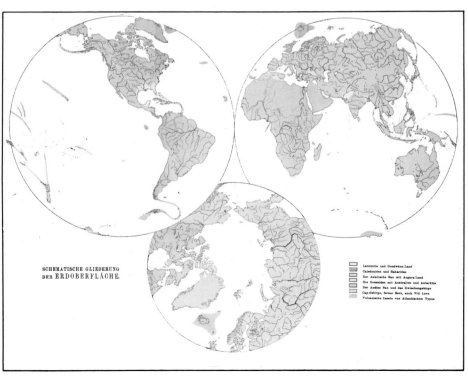

SCHEMATISCHE GLIEDERUNG DER ERDOBERFLÄCHE.

One of the earliest concepts of continental movement, illustrated by these maps, appears in an 1858 treatise by an American expatriate, Antonio Snider-Pellegrini. Unlike later theorists of leisurely continental drift, Snider-Pellegrini held that the lands bordering the Atlantic were blown apart in a cataclysm that resulted in the Biblical Deluge.

Africa. On the other hand, zoologists were surprised to find on the island of Madagascar, only 300 miles from the eastern coast of Africa, very few of the animals common to the nearby mainland. Instead, 34 of the 65 mammal species discovered on the island were lemurs, some of which were closely related to similar species observed in India, 3,000 miles to the northeast.

With the idea of a cooling earth in mind, the English ornithologist Philip Sclater postulated in 1864 that a land mass he called Lemuria once spanned the Indian Ocean, connecting Madagascar with India. As the crust contracted, he thought, the continent sank beneath the waves, leaving Madagascar as its only trace. Since lemurs are primitive ancestors of monkeys, Sclater concluded that Lemuria disappeared before higher forms of mammals appeared—an idea that gained credence from the evolutionary theories of Charles Darwin.

The Austrian paleontologist Melchior Neumayr was another proponent of lost continental links to account for the presence of identical forms of life on separate continents. In 1887, using the worldwide distribution of fossils as a guide, Neumayr reconstructed the topography of the earth as it might have appeared 190 million years ago. The map that emerged showed three huge supercontinents—one combining Greenland and North America, one consisting of Africa and South America, and one joining Australia and Asia—grouped around numerous smaller land masses in the area of what is now Europe. Neumayr's map inspired many other earth scientists to interpret their own findings in a similar fashion and construct roughly similar supercontinents.

By the end of the 19th Century, naturalists had discovered enough plant and animal fossils around the world to lend strong support to the belief that there must have been substantial land links for a considerable time between South America, Africa, India, Australia and Antarctica. Eduard Suess, in *The Face of the Earth,* proposed that there had been two supercontinents, which he called Gondwanaland—after the Indian kingdom where fossils typical of the whole area were first discovered—and Atlantis. Later, he speculated that there may have been as many as four original land masses. The most popular theory held that a system of wide land bridges, which have since collapsed, once linked the continents, enabling plant and animal life to move freely from one continent to another. It was generally agreed that the disappearance of these land bridges had been caused by the cooling and contracting of the planet.

The agreement was not universal. One of those who believed otherwise was Antonio Snider-Pellegrini, an American living in Paris, whose wide and eccentric interests ranged from ways of making the Sahara fertile to the meaning of comets. Snider-Pellegrini agreed that the earth had once been in

Bailey Willis' Mountain Machine

Geologist Bailey Willis

Throughout his career with the United States Geological Survey and, later, as a professor at Stanford University, geologist Bailey Willis was known for his daring hypotheses about the forces that shape the earth.

Like his mentor, John Wesley Powell, the director of the Geological Survey, Willis saw in uplifted and crumpled rock strata evidence of a dynamic globe, continually altered by stupendous forces. With Powell's encouragement, Willis attempted to reproduce the intricate folding of the Appalachian Mountains in laboratory experiments during the 1890s. Willis duplicated the many layers and varying thicknesses of the Appalachian strata in beeswax, and simulated their differing characteristics by making some of the strata more pliant with turpentine and stiffening others with plaster of paris.

A screw-driven piston provided the equivalent of mountain-building forces, compressing the miniature strata into a multitude of folds and fractures. Willis found that the resulting formations varied widely as he altered the arrangements and thicknesses of the stiff and pliant layers. From the variations, Wil-

lis drew detailed conclusions about the characteristics of the rock layers under the Appalachian ridges.

But a fundamental mystery remained: the source of the huge compressive forces that could deform solid rock. Toward the end of his career, Willis arrived at a bold theory that linked mountain building with the heat generated deep within the planet by radioactive decay. In 1938 he suggested that vast pockets of magma rising from these internal fires could cause crustal rocks to expand, resulting in lateral pressures strong enough to build mountains. Willis went on to guess—wrongly—that magma welling up beneath the Atlantic Ocean some 200 million years ago deformed the eastern coast of North America and uplifted the Appalachians.

In another bold, though erroneous, hypothesis, Willis attempted to account for the presence of similar life-forms on widely separated continents. If rising magma could build mountains on land, thought Willis, it might also have warped the sea floor, briefly uplifting land bridges across which living creatures might have migrated from continent to continent.

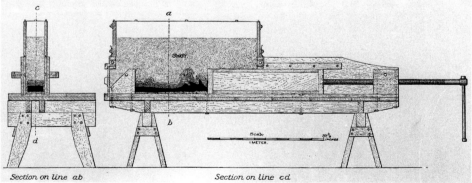

Miniature mountains rise in Bailey Willis' 1891 device for duplicating the compression and folding that transforms layers of sedimentary rock into mountains. The weight of loose lead shot tended to flatten the wax formations, simulating the effect of gravity.

Compressed by 300 pounds of lateral pressure, Willis' wax model of rock strata exhibits both the smooth folding and the abrupt faulting characteristic of the Appalachian Mountains.

An 1897 photograph of folded strata in the Appalachians bears witness to the power of mountain-building forces. Now arched and bowed, these layers of shale—formed by accumulated oceanic sediments—once lay flat and level on the floor of a prehistoric sea.

a molten state, but he parted company with those who held that continental features had remained fixed during the subsequent cooling and contraction.

In his book *La Création et Ses Mystères Dévoilés,* published in 1858, Snider-Pellegrini speculated that as the earth cooled, all the continents formed in a single mass on one side of the globe. He cited the coastal fit of Africa and South America as evidence, and drew a number of convincing maps to show how they might once have fitted. Snider-Pellegrini was thus the first scientist to suggest that the continents had moved long distances across the surface of the earth. Unfortunately, he made the same mistake that Alexander von Humboldt had made more than 50 years before, and went on to ascribe the splitting up of this single land mass to Noah's flood. By embracing this unfashionable catastrophic doctrine, he lost whatever scientific support his original notion might have generated, and his book was never taken seriously. Meanwhile, there was mounting evidence that the notion of sinking continents and land bridges should not be taken seriously either.

In 1735, the French scientist Pierre Bouguer sailed for South America as the joint leader of an expedition to the Andes. While making many other scientific observations, Bouguer calculated the height of the immense moun-

Geologist-explorer John Wesley Powell talks with a Paiute Indian chief in southern Utah (*left*) during a reconnaissance of the Grand Canyon in 1872. Powell concluded that a slow uplifting of the canyon, whose stratified walls are seen in an expedition sketch below, had occurred as the river eroded the bottom of the gorge.

tains, using the time-honored surveyors' method that began with establishing a vertical line with a plumb bob. Such calculations were customarily corrected for the assumed gravitational pull of the mountains, and Bouguer expected that the plumb line would be substantially deflected toward the great mass of the mountains. But the actual deflection was only a fraction of what he had foreseen, and after conducting such experiments on the slopes of Mount Chimborazo in what is now Ecuador, Bouguer noted in his journal that the interior must be virtually hollow, or certainly less dense than he imagined. It was, he marveled, almost as if the mountain were made of eggshells rather than of massive, solid rock that would exert a strong gravitational pull.

Bouguer was a brilliant scientist who was widely known in France (he had been royal professor of hydrography at the age of 15), and his findings in the Andes led to wild speculation in the French press about "hollow mountains." Scientific speculation was rather more sober but no less intense, particularly after measurements taken in the Himalayas during a British surveying expedition led by George Everest—in whose honor Mount Everest was named—produced similarly curious results.

In 1855, Sir George Airy, the British Astronomer Royal, attempted to explain this anomaly by suggesting that mountains had huge roots extending deep into the earth to support their weight; these roots, he said, were much less dense than the surrounding mantle and thus exerted a weaker gravitational pull. This explained the unexpected behavior of the plumb bob; its tendency to swing toward the mountaintop, which is more dense than air, was counteracted by its tendency to swing away from the mountain root, which is lighter than the mantle. Even more startling, Airy theorized that mountains, since they were composed of comparatively light granite, "floated" like enormous icebergs on the denser basalt layer of the earth's mantle. (Both granite and basalt are formed by the cooling of molten materials from the earth's interior, but basaltic crystals are finer and more tightly packed—and thus heavier.) "It is supposed," he said, "that the crust is floating in a state of equilibrium." Thus, as the mountains lost weight through inevitable erosion by wind and water, they rose gradually from the surface.

The concept of geological uplift gained some credence in the 1870s, after American geologist John Wesley Powell—a Civil War veteran who had lost his right arm at the Battle of Shiloh—had conducted a series of expeditions on the Colorado River and its tributaries. Traveling in small wooden dories down the Green River as it snakes its way southward through the towering rock canyons of the Uinta Mountains, Powell and his colleagues came to an extraordinary conclusion: The canyons—despite first appearances—had not been cut down from top to bottom by the river. Rather, the river had remained at a constant level while the surrounding land had risen slowly upward during millions of years; the river, carrying tons of abrasive mud, sand and gravel every day, had scoured each successive layer as the rock moved up against it. Had the mountains always been there, Powell reasoned, they would have diverted the river to a new course that presented less resistance. The "awesome abyss" of the Grand Canyon, said Powell, had been formed in the same way.

Clarence Dutton, who accompanied Powell on later expeditions to the Rocky Mountain region, expanded on the floating lands theme in 1889,

Thunderstorms drench the Grand Canyon in Thomas Moran's *Chasm of the Colorado,* completed in 1874. Moran, who painted the canyon at John Wesley Powell's invitation, tried to convey the awe he thought was lacking in Powell's geological descriptions.

27

when he published a revolutionary paper that was prosaically titled "On Some of the Greater Problems of Physical Geology." Dutton had spent 10 years studying the apparent uplifting, sinking, twisting and folding of the earth's crust on the plateaus of Utah, Arizona and New Mexico. He concluded that the equilibrium of the earth's crust is determined by density, that lighter materials rise to form continents, denser material sinks to form oceans. The granite continents ride higher than the heavier basalt of the sea floor in approximate hydrostatic balance. He called it the theory of isostasy, from Greek roots meaning "equipoise."

Dutton's theory was a disturbing concept for earth scientists who believed in sunken continents caused by a cooling and contracting planet. If his theory were proven, continents simply could *not* sink: It was physically and geologically impossible. In that case, there could be no denying that the "wrinkled apple" theory, around which so many geologists had built their careers and their reputations, was completely wrong. For his part, Clarence Dutton was forthright in expressing his scorn for his colleagues' cherished theory. "I dismiss it," he wrote, "with the remark that it is quantitatively insufficient and qualitatively inapplicable. It is an explanation which explains nothing we want to explain."

Furious scientific arguments followed in the wake of Dutton's pronouncements. Eduard Suess totally rejected the idea of floating continents and continued work on his scholarly opus, still convinced that global contraction determined the topography of the earth. The quest for alternative hypotheses produced some wild notions: In an article in the respected magazine *Nature*, the Reverend Osmond Fisher, an English rector, suggested that the match between Africa and South America resulted when the moon was ripped away from the earth, probably by the gravitational pull of a passing star. The Pacific basin, he said, was the scar left by this catastrophe, and as it began to fill with lava, what was left of the earth's crust broke up and slowly floated toward the depression. "This would make the Atlantic a great rent and explain the rude parallelism which exists between the contours of America and the Old World," wrote Fisher. (This imaginative idea about the origin of the moon was popular for many years, but in 1969 it would at last fail the ultimate test: Analysis of samples taken from the surface of the moon showed that it is in fact nearly four billion years older than either the Atlantic or the Pacific.)

Already challenged by the proponents of shifting continents and seabeds, the believers in a cooling earth were put to yet another test at the beginning of the 20th Century. Several experimenters—most notably the husband-and-wife team of Pierre and Marie Curie in France—demonstrated that certain elements in the earth, far from cooling off with the passage of time, are actually producing heat. These elements contain atoms of various weights, called isotopes, some of which are unstable and tend to decay until they reach a stable form. In this process, the isotopes release both heat and radioactive particles. Indeed, some scientists began to suggest that the entire world was in fact growing warmer. The earth sciences seemed clearly in need of an alternative to the theory of a cooling and contracting planet. And it would not be long in coming.

On December 29, 1908, a geologist named Frank B. Taylor presented an extraordinary paper to the Geological Society of America. Taylor, a Harvard dropout whose private studies had been financed in large part by his wealthy father, maintained that the continents had not sunk, but had instead shifted horizontally, in a "mighty creeping movement," to their present positions on the surface of the earth. He had based his ideas on a study of such mountain ranges as the Andes, Rockies, Alps and Himalayas, designated Tertiary mountains because they were evidently formed during the so-called Tertiary geological period, which began some 65 million years ago. It seemed apparent to Taylor that the configuration of these mountain ranges could have occurred only as a result of titanic lateral pressures that during long periods of time thrust the surface of the earth upward.

To explain how this could have taken place, Taylor proposed that a pair of protocontinents, one located over each of the earth's poles, had been ruptured by tidal forces—possibly when the moon, which Taylor believed was a comet "captured" by the earth, went into orbit around the planet. These great land masses then began creeping toward the Equator, and the Tertiary mountain ranges were thrown up as the moving continents collided with other land blocks that had remained stable. For the first time, he described the Mid-Atlantic Ridge, known even then as a major undersea mountain range running parallel with the coasts of Africa and South America, as the line of rifting between the two continents. The ridge, he surmised, "has remained unmoved while the two continents on opposite sides of it have crept away in nearly parallel and opposite directions."

Taylor's arguments for continental movement were the most cogent yet presented, but his ideas were either ignored or opposed by fellow scientists. As Bailey Willis, a professor of geology at Stanford University, declared in the same year: "The great ocean basins are permanent features of the Earth's surface and they have existed where they are now with moderate changes of outline since the waters first gathered." And then, 18 months later, a German meteorologist and explorer named Alfred Wegener put forward a hypothesis somewhat similar to Taylor's, which he called "continental displacement." This time the reaction of the scientific establishment was far different. Taylor had caused barely a ripple. Wegener would provoke a bitter and furious storm. Ω

Explorers' reports filtering back to Europe in the late 19th Century told of a strange valley that cut across the African landscape from the lower end of the Red Sea to Mozambique, 2,500 miles to the south. Scientists were puzzled because erosion could not have formed such a valley. The accounts told of a broad, flat valley floor bounded by steep, fractured walls often 2,000 feet high; of lakes trapped there with no outlet to the sea; and of territory seething with volcanic activity. Here was a geological enigma that was apparently produced by mysterious earth movements on a grand scale.

In 1893, John Walter Gregory, a Scotsman, became the first geologist to explore and name the Great Rift Valley. He soon agreed that water could not have carved the 30-mile-wide depression. Declared Gregory, "The Rift Valley rivers are due to the pre-existing valley, and have no more made it than a railway train has made the cutting through which it runs." Instead, Gregory concluded, subterranean stresses had fractured the earth along two parallel courses, and the intervening strip of land had subsided.

Symptoms of geological unrest, such as earthquakes, lava flows and mineral-laden hot springs, confirm that the faulting and subsidence that opened Africa's Rift Valley 20 million years ago continue today. And geologists now believe that the same forces that widen the valley by about a millimeter per year have shaped most of the earth's features.

A gash on the face of Africa, the most dramatic portion of the Great Rift Valley stretches 1,500 miles, from the Red Sea to Tanzania (*map*). At Hell's Gate in Kenya (*left*), the 600-foot walls of the Rift bear scars of the faulting that lowered the valley floor.

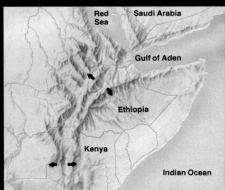

33

A dormant crater on the slopes of the 9,100-foot Longonot volcano overlooks Africa's Rift Valley in Kenya, where geologist John Walter Gregory first glimpsed the geological wonder in 1893. Longonot has not erupted for a century, but active volcanoes dot the valley.

Volcanic formations near Hell's Gate testify to the geological violence that abounds in the Rift Valley. These massed basalt columns were formed when lava fractured in a regular pattern as it cooled and hardened.

Its waters colored a lurid red by algae, the soda-encrusted Lake Natron sprawls across 400 square miles of the Rift Valley floor on the Kenya-Tanzania border. Whorls of drying soda on the lake surface mark submerged hot springs, where water laden with the caustic mineral wells up into the lake.

Salt stained by iron compounds forms a
multicolored cascade in the Danakil Depression,
a low-lying wasteland at the northern end
of the Rift Valley. The salty waters of the area's
many volcanic springs leave such formations
behind as they evaporate in the desert heat.

Salt flats surround a puddle of brine in the
Danakil Depression. Once an arm of the Red Sea,
the Danakil dried and became a vast salt flat
when the earth movements that created the Rift
Valley separated the region from the sea.

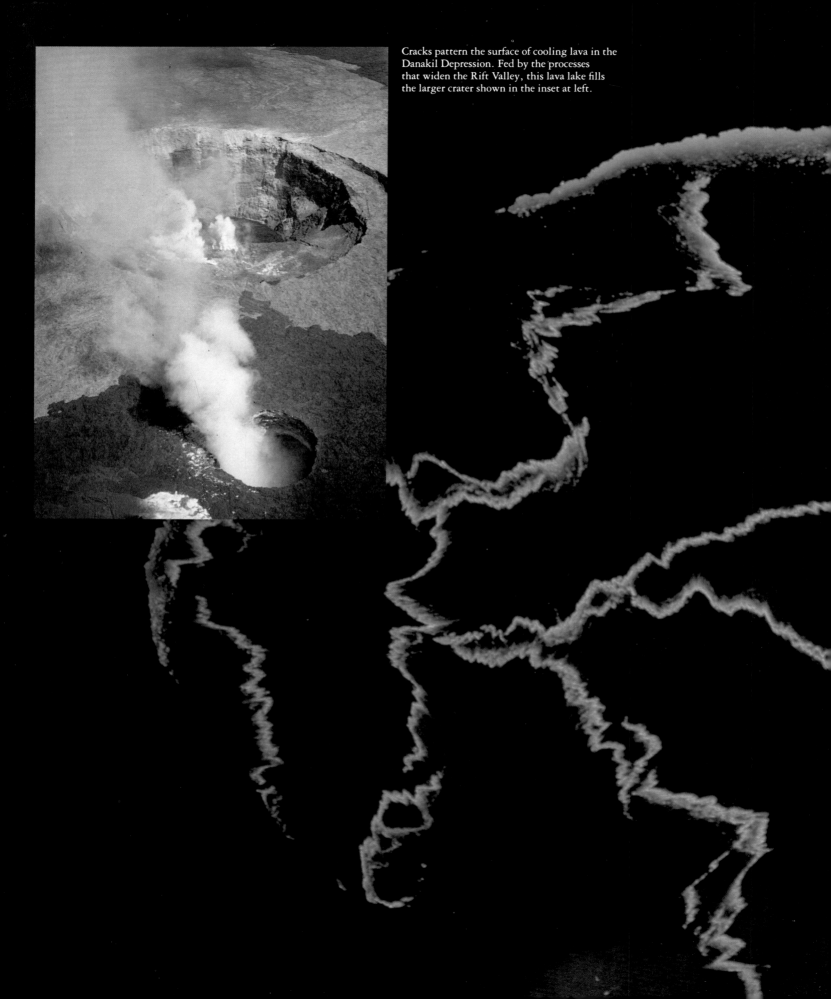

Cracks pattern the surface of cooling lava in the Danakil Depression. Fed by the processes that widen the Rift Valley, this lava lake fills the larger crater shown in the inset at left.

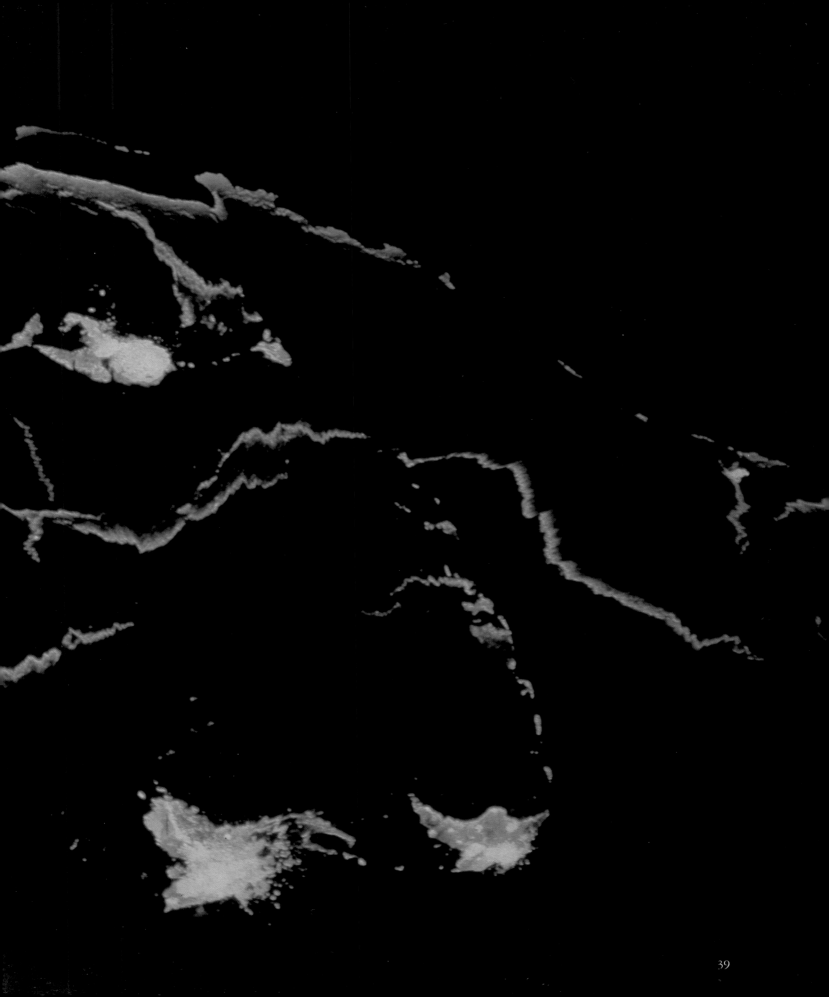

Chapter 2

THE "IMPOSSIBLE HYPOTHESIS"

One day in the autumn of 1911, Alfred Wegener, a 31-year-old lecturer in astronomy and meteorology, was browsing in the library at the University of Marburg in western Germany when, quite by chance, he picked up a scientific paper advancing the theory that a land bridge had once existed between Brazil and Africa. Although the idea of former connections between now-distant continents was familiar to Wegener, the evidence presented in this paper—descriptions of identical fossils of plant and animal forms that could not possibly have crossed an ocean and yet were found on opposite sides of the Atlantic—came as a surprise. The information so intrigued him that he began to comb other research papers for more details, and before he knew it, what had started as a momentary digression during a cursory tour of the library shelves had become an obsession. It was, as it turned out, a fateful event for 20th Century science.

As Wegener learned more about the existence of creatures as diverse as monkeys and snails, of plants as varied as ferns and heather, in lands as widely separated as Africa and South America or Europe and North America, he was seized again by an earlier speculation. Like many scientists before him, Wegener had noted the apparent fit between the Atlantic coastlines of Africa and South America, and had briefly wondered whether the continents could have been a single land mass at one time. But the idea that they could have moved the distance represented by the breadth of the Atlantic Ocean seemed so fantastic that he had dismissed it as unworthy of further consideration. It now occurred to him that the evidence he was seeing, far from proving the existence of land bridges, seemed to confirm instead his fleeting intuition that the continents had once been united and had since drifted apart. As he later wrote, "A conviction of the fundamental soundness of the idea took root in my mind."

How Wegener took hold of that conviction, how he embellished it and pruned it, scrutinized it and stood by it, and the tortuous route by which it won eventual recognition after sinking into near-oblivion constitutes one of the most remarkable stories in the annals of science. Without a specialist's initiation into the arcane minutiae of the fields he was dabbling in, acting solely on what had begun as a flash of insight, Wegener was to work out a comprehensive theory that would completely alter the way geologists, geophysicists and paleontologists view the earth and its evolution, and that would broaden and enrich the fields of biology, zoology and oceanography. It would not gain that eminence, however, before making Wegener the object of derision among scientists the world over.

A massive iceberg rises from a calm sea off the coast of Greenland, the wilderness to which Alfred Wegener journeyed on three scientific expeditions between 1906 and 1930. The stately migrations of Greenland icebergs are sometimes credited with inspiring Wegener's remarkable theory of continental drift.

Alfred Wegener was born in Berlin on November 1, 1880, the youngest child of an evangelical preacher. By the time he was in his teens, he had developed a marked interest in the earth sciences and a deep yearning to visit Greenland, where the most exciting and advanced studies of geophysics were then being made. Determined to get to that harsh clime one day, he made it a point throughout his years of schooling to build and test his physical endurance with exhausting hikes, days-long skating expeditions, perilous mountain climbing and enthusiastic skiing. He studied astronomy at the University of Berlin, where he received a doctoral degree in 1904. Shortly thereafter, he published his first scientific paper, a treatise on the history and use of the Alphonsine tables—a compilation of planetary motions that had served astronomers for 400 years after their completion in 1252 under the patronage of King Alfonso X of Castile. Wegener's professional career was well launched; the paper was widely read and highly regarded by German scholars.

Meanwhile, Wegener had become entranced with the infant science of meteorology, which was then being nurtured by rapid improvements in communications technology, especially telegraphy. The increased capability to track the progression of weather systems was yielding new knowledge—and fascinating new questions—with breathtaking speed. Wegener determined that the most advanced work on meteorology in Germany was being done at the Prussian Aeronautical Observatory in Tegel; he took a position there immediately after his graduation from the University of Berlin, and soon pioneered the use of balloons to determine the path of air currents. His studies in astronomy and meteorology were to be the basis of his professional credentials for the remainder of his life. But instead of limiting him, as specialization limited the view of so many of his colleagues, the disciplines served Wegener as springboards from which his nimble mind ranged over all the fields encompassed by geophysics—climatology, volcanology, magnetism, oceanography, hydrography, glaciology.

He had been out of the university about two years when the opportunity came to realize his boyhood dream of exploring Greenland; he was invited to join a 1906 Danish expedition to the island as official meteorologist. He accepted with alacrity, and for the next two years lived, traveled—and exulted—in the most rigorous conditions imaginable. "We feel like shock troops of humanity in the battle with the deadly powers of nature!" he wrote. "Science against the icy blasts of snow!" The expedition mapped the Greenland coast—which until then had not been fully explored—and, in addition to conducting atmospheric research, made a number of lunar observations used to determine longitude. Among the many reports that Wegener produced was one noting that the longitudinal calculations of his party differed from earlier measurements made in 1823 and 1870. The discrepancy lodged in Wegener's mind, and in time he found a use for it.

On his return to Germany in 1908, Wegener reentered the world of academe by taking a job as a private tutor at the Physical Science Institute at the University of Marburg. There he exhibited a talent for expounding complex subjects with casual ease. Combined with the force of his personality, the clarity of Wegener's vision inspired great enthusiasm and fierce loyalty among his students. "I think they would have gone through fire for Wegener," recalled fellow professor Hans Benndorf, who was himself mightily impressed by the young instructor's mind: "He acquired his knowledge mainly by intuitive means, never or only quite rarely by deduction from a formula. If matters concerning physics were involved, that is, in a field distant from his own field of expertise, I was often astonished by the soundness of his judgment. With what ease he found his way through the most complicated work of the theoreticians, with what feeling for the important point! He would often, after a long pause for reflection, say 'I believe such and such' and most times he was right, as we would establish several days later after rigorous analysis. Wegener possessed a sense for the significant that seldom erred."

With a frankness that charmed his students and disarmed his associates, Wegener often confessed to having no particular gift for mathematics, and expressed considerable impatience with "such mathematical treatises as I cannot understand." Impenetrable arguments, he suggested, might simply be wrong: "When one cannot follow the printed or written word one should not always put the blame on oneself." Wegener's refreshing approach made him increasingly popular; soon senior professors were dropping in to listen to their young colleague. After three years of teaching, Wegener collected his lectures in a volume titled *The Thermodynamics of the Atmosphere,* and he proved to be as lucid in print as in the classroom; the book soon became a standard text for meteorological studies all over Germany. His textbook had just been published when Wegener had his fateful library encounter with the literature on fossils and land bridges.

The idea that strips of land had once connected the continents and had later sunk into the sea was an outgrowth of, and depended on, another widely accepted theory—that the earth was cooling and contracting. As the core had cooled and shrunk, the argument went, the outer crust had collapsed inward; the mountains had risen and oceans had formed in the depressions, covering the erstwhile land bridges.

Wegener, like many other scientists of his time, perceived serious flaws in this view of the earth's development. An obvious and longstanding problem was that if mountains had come into existence as a result of the earth's contraction, they should be evenly distributed over the globe. Instead, as reference to any map clearly showed, they occur in narrow curvilinear belts, more often than not on the edges of continents. Another contradiction was presented by the age of the mountains; a cooling earth would have formed them more or less simultaneously, but scientists knew from fossil evidence and from recently developed (and still only approximate) radioactive dating techniques that the Appalachians, for instance, are millions of years older than the Rockies and that the Caledonian mountain system in Great Britain is far older than the Alps.

Still another flaw in the contracting-earth theory came to light when gravity measurements showed that the density of continental crust is distinctly different from that of oceanic crust. The continents are composed of relatively lightweight rocks—in Wegener's day called sal or sial, a contraction of the names of its principal constituents, silicon and aluminum. Oceanic crust, then deemed to be the earth's mantle, is heavier; it was called sima, for its content of silica and magnesium.

The knowledge of this fundamental difference between continents and sea floors gave rise to the theory of isostatic balance—that the continents float in the denser underlying mantle as icebergs rest in the sea, with a submerged mass far larger than the exposed portion. (Modern theories of the earth's crust have dispensed with the terms "sial" and "sima," and take into account a number of variegated layers between the earth's outer crust and its core, but the fundamental principle—that continental and oceanic crusts differ in composition and density and exist in isostatic balance—still holds.) Even if the light crustal rocks of the vanished land bridges had somehow been forced down into the denser sea bottom, they would have had to rise again, just as a buoyant object dropped into water must bob back to the surface. Icebergs do not sink, they break up and drift apart; continents, by Wegener's reasoning, could be assumed to do the same.

Just a few months after his fortuitous discovery in the Marburg library, Wegener had worked out his extraordinary thesis. On January 6, 1912, he stood before a meeting of the august Geological Association in Frankfurt and delivered an address titled "The Formation of the Major Features of the Earth's Crust (Continents and Oceans)." Four nights later he gave a similar lecture to the Society for the Advancement of Natural Science in Marburg. In both lectures, Wegener boldly challenged the prevailing theories of sunken land bridges, and suggested instead that continents had once been united and had subsequently broken apart and drifted to their present positions. Neither address was well received; indeed, some members were openly indignant that such a ludicrous notion should be aired in their presence.

Wegener did manage to win one staunch ally—Professor Wladimir Köppen, Director of Meteorological Research at the German Naval Observatory in Hamburg, affectionately known throughout Germany as "The Grand Old Man of Meteorology." Köppen, who was to become a lifelong colleague and collaborator (and Wegener's father-in-law), would later write that many a scientist had wondered about the resemblances between coastlines on opposite sides of the Atlantic, but "now this similarity had been noticed by an expert geophysicist, a brilliant man of unbounded energy, who would spare no pains in following up the matter and gaining any facts from other fields of science that might seem to have a bearing on the question."

Köppen also observed that the budding scientist had taken a considerable risk in straying from his meteorological research and encroaching on other sciences: "To work at subjects which fall outside the traditionally defined bounds of a science naturally exposes one to being regarded with mistrust." But the theory of continental drift already held such a grip on Wegener's mind that he could not abandon it.

He did postpone further work on it, however. In the spring of 1912 he joined another expedition to Greenland, where he made a 750-mile journey—the longest traverse of the icecap ever undertaken—across the island's widest point. The expedition was the first to winter on the glacier, and it produced volumes of observations of interest to both meteorologists and glaciologists. Soon after his return to Germany in 1913, he married Köppen's daughter Else, only to be drafted into the German Army as an officer when World War I broke out a scant year later. He was a far-from-enthusiastic recruit, not because of the threat of hardship and danger (which rather appealed to his nature), but because he profoundly hated war. He was to have a relatively merciful escape, however; a minor wound in the arm and then another in his neck ended his combat duty, and he was

assigned to the Army's weather forecasting service, where he again found the time to pursue his scientific interests.

By 1915, Wegener had drawn various strands of evidence from several scientific disciplines into a tightly woven exposition of his theory of continental drift titled *The Origin of Continents and Oceans*. He made no apologies for being an interloper in other sciences. As his brother Kurt later explained, "This book was concerned with reestablishing the connection between geophysics on the one hand and geography and geology on the other, a connection which had been completely broken by specialist development of these branches of science." The slim 94-page volume was a tour de force, the first comprehensive attempt to explain the evolution of the earth's geography. But 1915 was not a propitious time for launching a new academic theory. With public attention focused on the War, *The Origin of Continents and Oceans* had a small circulation and attracted little notice. Three more editions followed in 1920, 1922 and 1929, each adding further elaboration to its predecessors.

Wegener's treatise began by establishing the difference between the two layers of the earth's crust. The outermost layer—the lighter, granitic rocks then called sial—comprised the continental material. He likened sima, the material of the sea floor, to pitch, which breaks like a brittle solid when hit with a hammer but flows plastically under its own weight in the course of time. If the continents were floating in this pitchlike substance, he argued, and if, as the theory of isostasy held, they were moving vertically to achieve equilibrium, then there was nothing to stop them from moving horizontally as well. The fossil evidence clearly indicated the past existence of land connections between continents; inasmuch as the concept of sunken land bridges could not be defended simultaneously with the theory of isostasy, then only one alternative remained—the continents had once been joined together and had since drifted apart.

After demolishing the theory of sunken land bridges and establishing the essential differences between sial and sima, Wegener proposed that the geological boundaries of the continents lay not on their shorelines, but at the line of demarcation between the two kinds of material—the edges of the continental shelves. When Wegener fitted Africa and South America together along their continental shelves, he found that large blocks of ancient rock called cratons formed contiguous patterns across the dividing line. The mountains that run from east to west across South Africa seemed to link with the range near Buenos Aires in Argentina; the distinctive rock strata of the Karroo system in South Africa—layers of sandstone, shale and clay laced with seams of coal—were identical to those of the Santa Catarina system in Brazil. "It is just as if we were to refit the torn pieces of a newspaper by matching their edges," he wrote, "and then check whether the lines of print run smoothly across. If they do, there is nothing left but to conclude that the pieces were in fact joined in this way. If only one line was available for the test, we would still have found a high probability for the accuracy of fit, but if we have *n* lines, this probability is raised to the *n*th power."

It had occurred to Wegener in the early 1920s that more evidence to support his thesis could be found by tracing ancient changes in climate. With the collaboration of his father-in-law, who brought to bear his extensive meteorological expertise, Wegener plotted the worldwide distribution

Stained and pitted by marine organisms during centuries underwater, the columns of the Second Century B.C. Temple of Jupiter Serapis jut from shallow water in Pozzuoli, Italy.

46

Even as Alfred Wegener's contemporaries derided his notion that continents could move, they must have known about the dramatic evidence of a dynamic earth provided by the small town of Pozzuoli, Italy. Indeed, as far back as the 1830s, Sir Charles Lyell, the famous British geologist, had described in detail the remarkable phenomena affecting the village near the Bay of Naples.

As Lyell noted, Pozzuoli continually rises and falls in a series of centuries-long undulations so gentle that the slender columns of the 2,100-year-old Temple of Jupiter Serapis still stand, providing a graphic record of the land's slow fluctuation. Built on a low-lying area near the shore, the temple began sinking soon after its construction, until it rested in 23 feet of water. Around 1000 A.D., the process reversed and the land began a long, sometimes dramatic rise: In 1538, the town was lofted 22 feet in 48 hours. Today, after a period of subsidence, the temple stands in the shallows, but evidence of its watery past is etched high on the marble columns in banded scars left by burrowing shellfish.

Scientists have located the cause of Pozzuoli's movement in deep subterranean layers of porous rock. There, trapped groundwater is superheated by volcanic activity, generating enormous pressure. When the pressure peaks, the ground bearing Pozzuoli rises; when it diminishes, the town gently settles.

of fossils and rock types, which indicated the former locations of tropical jungles, deserts and icecaps.

The history of climate changes that emerged could only be explained by rearranging and uniting the continents. He discovered, for instance, that the island of Spitsbergen, which has a polar climate and lies blanketed in ice, offers fossil evidence of former stands of beech, poplar, oak, maple and elm, which can thrive only in temperate climates. Still older remains of tropical tree ferns and sago palms are also found there. Such changes in vegetation, said Wegener, could not be explained unless Spitsbergen itself had moved from the tropics to the Arctic Circle. In other places all over the world, Wegener found similar evidence of changing climates provided not only by plant fossils but by animal remains as well. "It is probably not an exaggeration to say," Wegener wrote, "that if we do not accept the idea of former land connections, the whole evolution of life on earth must remain an insoluble riddle."

By the time he published the third edition of his book in 1922, Wegener had contrived to join not only Africa and South America, as he had done in the first edition, but all the continents into one huge land mass. He cited eight major geological features that matched perfectly in the reconstruction: They included the diamond fields of South Africa and Brazil; the coal fields of Britain, Belgium and the Appalachian Mountains; the thick band of red sandstone extending from the Baltic across Norway and Britain to Greenland and North America; and—an unfortunate inclusion—the edge of the ice sheet that was known to have covered Europe and North America two million years ago. In Wegener's reconstruction, the edge of the ice sheet in Europe was joined "without a break" to that of North America; this was a mistake that was shortly to be compounded.

Wegener called this primordial continent Pangaea, from Greek roots meaning "all land," and calculated that it existed 300 million years ago. He thought that Pangaea began to separate into northern and southern proto-continents around 200 million years ago, and that the resulting land masses slowly pushed their way through the viscous sea floor on their long journeys to their present positions.

Wegener's book offered a new explanation for the rise of mountain chains that was to prove far more tenable than the one based on the contracting-earth theory. When the leading edge of a drifting continent encountered resistance, he said, it crumpled and folded into great mountains. As America had drifted westward, for example, with the Atlantic opening in its wake, the resistance of the cooled bedrock of the ancestral Pacific had caused the folding of the Rockies and the Andes. Similarly, India had drifted northward after separating from Antarctica, then collided with Asia and thereby uplifted the Himalayas.

Citing as evidence the longitudinal observations taken during his first expedition to Greenland, which indicated that Greenland was west of its position as recorded in 1823 and 1870, Wegener asserted that drift was still in progress—at an astounding velocity of 118 feet per year (years later, scientists would place the figure at less than one inch per year). The error was the fault of imprecise methods of measurement and not the theory, but combined with his reconstruction of the two-million-year-old ice sheet, it led Wegener seriously astray in his scenario for the movements of North America and Europe. He concluded that the two continents separated from each other and from Greenland only one million years ago—hundreds of millions of years after the shaping of the other continents—and had since sped to their present-day positions. Such a timetable could hardly convince Wegener's contemporaries—most of whom doubted that continents could move at all, let alone at such blazing speed—of the soundness of his thinking.

Wegener knew that if his startling hypothesis was to be at all credible it must go beyond the conclusion that continents move: It must also explain what moved them. He made the attempt by postulating two possible sources for the enormous amount of energy required: centrifugal force—which tends to move rotating objects away from the center of rotation—generated by the spinning of the earth; and the gravitational attraction of the sun and moon, which causes the ebb and flow of tides. Wegener suggested that the centrifugal force pushed continents away from the Poles toward the Equator, and that on an eastward-turning earth the lunar-solar drag would tend to move the continents in the opposite direction. But Wegener knew all too well that these suggestions left a great many questions unanswered. "It is probable," he admitted, "that the complete solution of the problem of the driving forces will still be a long time coming."

Few scientists other than the German-speaking scholars of his own country paid much heed to Wegener during the seven years that elapsed between the publication of the first edition of *The Origin of Continents and Oceans* in 1915 and the appearance of the third edition in 1922. But with the translation of the third edition into English, French, Swedish, Spanish and Russian, his ideas were put into international circulation. And when scientists the world over perceived the challenge to the fundamental principles of the earth sciences implicit in Wegener's bold theory, they were quick to respond; many of them turned on the German meteorologist with savage fury.

Members of England's Royal Geographical Society took the subject of continental drift under consideration at a January 1923 meeting in London. While they generally agreed that the theory offered convenient explanations of much that was still puzzling about the earth, they completely rejected it. One geologist pointed out that the contracting-earth theory was so universally accepted that no one who "valued his reputation for scientific sanity" would dare advocate an extraordinary theory like continental drift. Another described Wegener's views as "vulnerable in almost every statement."

A geologist named Philip Lake delivered the most blistering attack, not only on the theory as such, but on its author. "Wegener is not seeking the truth," said Lake, "he is advocating a cause and is blind to every fact and argument that tells against it." He accused the German scientist of stretching, contorting and twisting the continents in a misbegotten effort to fit them together. "It is easy to fit the pieces of a puzzle together if you distort their shapes," Lake sneered, "but when you have done so your success is no proof that you have placed them in their original positions. It is not even proof that the pieces belong to the same puzzle, or that all the pieces are present." An even swifter and more summary indictment awaited Wegener in America, where the president of the prestigious American Philosophical

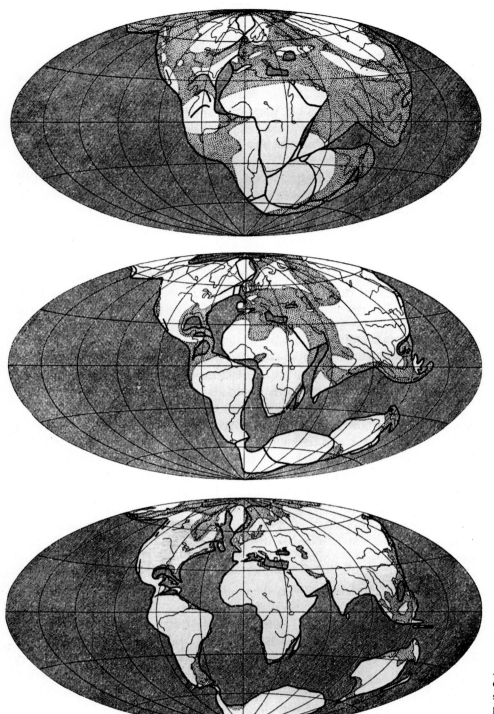

An illustration from Alfred Wegener's *Origin of Continents and Oceans* shows how he thought the supercontinent Pangaea split to form the present-day continents. The stippled regions denote continental areas covered by shallow seas.

Society in Philadelphia pronounced the idea of continental drift to be "utter, damned, rot!"

Wegener received a further mauling in 1924, when the British astronomer and geophysicist Harold Jeffreys published a treatise titled *The Earth, Its Origins, History and Physical Constitution*. In a few sentences Jeffreys contemptuously dismissed the geological and biological evidence advanced by Wegener and proceeded to attack the theory at its weakest point—its reliance on global rotation and gravitational pull for the driving energy of continental drift. With a few simple calculations, he demonstrated that the crust of the earth was far too strong to be affected by those forces. He also pointed out, with devastating effect, that a gravitational

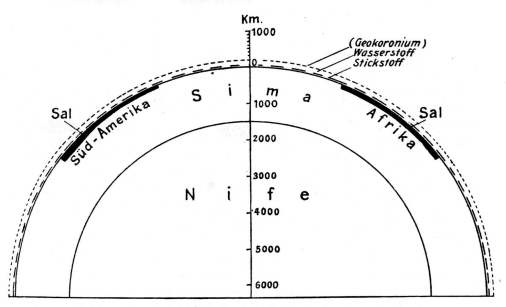

Wegener used this cross section of the earth, from the core (Nife) to the outer atmosphere (Geokoronium), to illustrate his theory that the continents are composed of a relatively light material (sal) and float like vast rafts on a layer of denser rock (sima).

attraction strong enough to shift continents would also stop the rotation of the earth in less than a year. He asserted categorically that there was no force capable of moving continents; if the force did not exist, he argued, then continents did not move. In sum, he wrote, Wegener's idea was "an impossible hypothesis."

Such attacks took their toll on Wegener's career. One companion of that time recalled the "depressing" days when Wegener "had to argue with his opponents or even defend himself against apparent misunderstandings." Despite his undisputed talents as a teacher, and the continuing loyalty of his close associates, Wegener remained a mere lecturer and was unable to obtain a professorship in a German university. "One heard time and again," a colleague remembered, "that he had been turned down for a certain chair because he was interested in matters that lay outside its terms of reference." Wegener never recorded his feelings about being thus rejected, but in 1924 he left Germany for the University of Graz in Austria, where a sympathetic administration created a chair of meteorology and geophysics especially for him. There, he was able to combine orthodox meteorological pursuits with further delving into his theory of continental drift. He also found his new associates a good deal more responsive to his ideas.

Despite their general rejection of the theory of continental drift, scientists somehow could not quite lay it to rest. In November of 1928, Wegener was invited to New York to attend an international symposium sponsored by the American Association of Petroleum Geologists. He eagerly accepted the chance to explain his views, only to find that the few voices of support raised at the meeting were quickly drowned out by a chorus of hostile dissenters, who criticized not only his hypothesis but his scientific credentials as well. One after another, delegates to the symposium stood up to express, with crushing sarcasm, grave doubts about the possibility of continental drift. Some barely troubled to justify their rejection of the hypothesis; others demonstrated errors of detail and used them to discredit the whole theory; a few seemed unable to restrain their anger that the idea was being seriously considered at all.

Professor Rollin T. Chamberlin of the University of Chicago attacked Wegener's geological evidence on 18 separate counts, claiming it ranged from unlikely to ludicrous. "Wegener's hypothesis in general," he said, "is of the footloose type, in that it takes considerable liberty with our globe, and is less bound by restrictions or tied down by awkward, ugly facts than most of its rival theories."

A professor of paleontology at Yale University, Charles Schuchert, provoked much hilarity by displaying pictures of a globe on which he had elaborately tried, and spectacularly failed, to fit together obviously incongruent coastlines such as those of North and South America. He also pointed out that erosion would have substantially altered the shape of the coastlines over long periods of time, yet Wegener was suggesting, by matching Africa and South America, that the fracture line had retained its shape for 120 million years. "Is there a geologist anywhere," asked Schuchert, "who will subscribe to this startling assumption?"

Professor Bailey Willis of Stanford University picked up on the same theme, charging that Wegener's supposed fit of the continental coastlines was illusory. If continents were drifting through a layer of the earth's crust, said Willis, the stresses of the movement would utterly destroy the original configurations; the apparent fit of Africa and South America could therefore be nothing more than coincidence. William Bowie of the United States Coast and Geodetic Survey used the nagging question of the driving force as the basis for his attack. If the continents were being propelled toward the Equator by some mysterious force, as Wegener had suggested, then how, Bowie asked, could four of the seven continents remain concentrated in the Northern Hemisphere, three of those on one side of the earth? Of a total of 14 speakers, hardly anyone had a favorable word for the idea of continental drift. One scientist who wrote about the symposium may unintentionally have accounted for much of the animosity when he complained, "If we are to believe Wegener's hypothesis, we must forget everything which has been learned in the last 70 years and start all over again."

Wegener himself spoke only briefly and said little in his own defense. Perhaps he had heard too many attacks to know where to start defending himself; perhaps he was so serenely convinced of the validity of his hypothesis that he saw nothing to be gained by arguing about details. Whatever the reason, he listened intently but silently throughout the symposium, smoking his pipe, to all appearances unmoved by the barrage of criticism.

On his return to Germany he went right ahead with a fourth and final edition of *The Origin of Continents and Oceans*, although this time he acknowledged the difficulties of trying to answer his critics. "Scientists still do not appear to understand sufficiently that all earth sciences must contribute evidence toward unveiling the state of our planet in earlier times, and that the truth of the matter can only be reached by combining all this evidence," he wrote. "We are like a judge confronted by a defendant who declines to answer, and we must determine the truth from circumstantial evidence. All the proofs we can muster have the deceptive character of this type of evidence. How would we assess a judge who based his decision on part of the available data only?

"It is only by combining the information furnished by all the earth sciences that we can hope to determine 'truth' here, that is to say, to find the picture that sets out all the known facts in the best arrangement and that therefore has the highest degree of probability. Further, we have to be prepared always for the possibility that each new discovery, no matter what science furnishes it, may modify the conclusions we draw."

Wegener also included in the fourth edition a few tidbits of supporting evidence offered by other scientists around the world. The respected Harvard geologist Reginald A. Daly had by then accepted Wegener's hypothesis but with a proviso: Daly proposed that the spinning planet developed bulges in certain places, and that the continents, instead of plowing through the earth's crust as Wegener maintained, might have slid downhill in a kind of global landslide. And the British geologist E. B. Bailey favored Wegener's theory because it accounted for the geological similarities between the Caledonian mountains in Ireland, Scotland and Norway, and the Appalachians on the opposite side of the Atlantic.

In Switzerland, Wegener had a distinguished and vigorous ally in Émile Argand, founder of the Geological Institute at Neuchâtel. A versatile genius who frequently worked for days without sleep, Argand had become

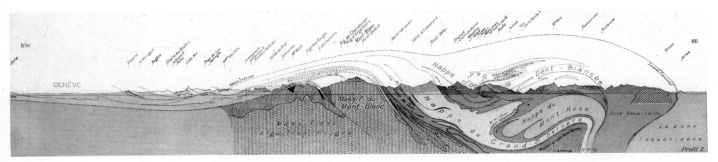

A 1911 drawing by Swiss geologist Émile Argand diagrams the buckled strata beneath Mont Blanc and surrounding peaks in the French and Italian Alps. Argand came to believe that only the collision of drifting continents could deform the earth so extensively.

interested in geology in 1905, at the age of 26. The Alps held a particular fascination for him, and he embarked on an intensive study of the complex geological events that had produced the great mountain chain.

At first, Argand accepted the traditional explanation that titanic forces generated by a cooling and contracting earth had thrust the mountains upward. But when he read the first edition of Wegener's work, in 1915, he promptly changed his mind, concluding that continental drift was the key to the formation of the range. In November 1916, Argand expounded on Wegener's theories at a meeting of the Neuchâtel Society of Natural Sciences. His listeners were startled both by Argand's departure from geological orthodoxy and by his political boldness: Anti-German bias was widespread in neutral Switzerland at the time, and citizens were being enjoined not to read materials published in Germany, much less lecture about them. But Argand persevered and was soon applying Wegener's theories to mountain ranges throughout the world. He explained his work to the International Geological Congress held in Brussels in the summer of 1922, thoroughly impressing his audience with the power of his intellect—but gaining few converts.

Wegener's most compelling and enthusiastic support came from a South African geologist by the name of Alexander Du Toit. South African scientists were far more favorably disposed to the idea of continental drift than their colleagues elsewhere for a straightforward reason: All around them the South Africans could see a plethora of geological phenomena that closely resembled those of the other continents in the Southern Hemisphere—as Du Toit could testify from personal experience. After spending five months in Brazil, Uruguay and Argentina, he wrote that he had "great difficulty in realizing that this was another continent." Not only did he find the same plant and animal fossils he knew at home, but he found them in the same complex sequence, embedded layer by layer in the rock. Similarities between the two sides of the ocean, he claimed, "are now

Folded limestone forms an S-shaped cliff near
Grenoble, France, typifying the Alpine crustal
deformations that helped convince Émile
Argand of the reality of continental drift.

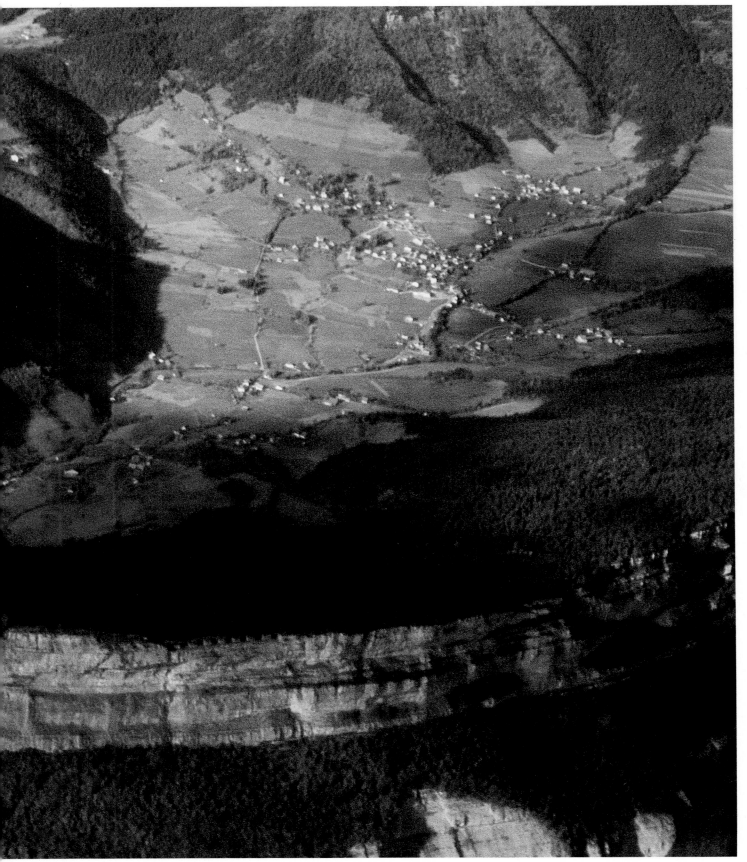

South African geologist Alexander Du Toit—one of the few scientists to advocate continental drift during Alfred Wegener's lifetime—prepared this map to illustrate his belief in the existence of two ancient supercontinents separated by a narrow sea.

known in such numbers that it is no longer possible to imagine them accidentally co-existing." Since scientists had already determined that it takes only a few miles of water to stop the spread of a diversified plant regime, Du Toit was confident he had conclusive proof that the continents were once joined.

However, in a book dedicated to Wegener and titled *Our Wandering Continents,* Du Toit proposed a prior configuration for the continents that was different from Wegener's. Instead of a single supercontinent, Du Toit reconstructed the southern continents at the South Pole, and grouped the northern continents near the Equator. Borrowing from Eduard Suess, the 19th Century Austrian geologist, he called his southern supercontinent Gondwanaland; he dubbed his northern land mass Laurasia, a coinage combining Laurentia—Suess's name for Greenland and much of North America—and Asia. Du Toit devoted most of his book to Gondwanaland, and as evidence for its existence he produced an impressive mass of data far more detailed than anything Wegener had attempted.

Under normal circumstances, such a dossier would have been convincing proof of the existence of Gondwanaland, if not of the validity of the whole theory of continental drift. But Du Toit's style was too flamboyant for his more sober colleagues. In a typical passage, he railed against "the dumbfounding spectacle of the present continental masses, firmly anchored to a plastic foundation yet remaining fixed in space; set thousands of kilometers apart, it may be, yet behaving in almost identical fashion from epoch to epoch and stage to stage like soldiers at drill; widely stretched in some quarters at various times and astoundingly compressed in others, yet retaining their general shapes, positions and orientations; remote from one another throughout history, yet showing in their fossil remains common or allied forms of terrestrial life; possessed during certain epochs of climates that may have ranged from glacial to torrid or pluvial to arid, though contrary to meteorological principles when their existing geographic positions are considered—to mention but a few such paradoxes!" His exuberant overdramatizations succeeded only in devaluing his substantial contributions to the evidence. "This," sniffed an academic critic, "is the colorful language of a pamphleteer."

Many scientists tried later to explain why the reaction to Wegener's theory was so predominantly hostile. Some argued that he was simply premature, too far ahead of his time to be accepted. Some suggested that any scientific hypothesis that contradicts the prevailing view will be set aside at least for a time in the hope that it will turn out to be invalid. But Wegener was not just set aside: His ideas were ridiculed and relegated to the realm of fantasy. Many of his colleagues preferred to close their minds to obvious contradictions in the prevailing theories about the origin of the earth, rather than consider Wegener's proposals. Paleontologists, for example, clung to the idea of sunken land bridges long after geophysicists had proved that the sinking of such bridges was impossible; and geophysicists themselves rejected Wegener on the grounds that he had not explained how the continents could have moved through the solid crust of the ocean floors.

In the final analysis, Wegener may have been rejected because, as Professor Köppen noted, he was an outsider. He was neither a geologist, nor a paleontologist, nor a biologist, yet his hypothesis trampled across all those fields, crushing underfoot long-held and cherished notions. "The trouble must partly have been," Oxford geologist Anthony Hallam would write four decades later, "that he was not an accredited member of the professional geologists' club. We of course now see it as a positive advantage that Wegener had not been brainwashed by the conventional geological wisdom as a student. His position was an advantage because he had no stake in preserving the conventional viewpoint. He was not an amateur, but an interdisciplinary investigator of talent and vision who surely qualifies for a niche in the pantheon of great scientists."

Wegener held steadfastly to his theory, in the face of public vilification and friendly support alike, with the certainty of a visionary. He continued to teach astronomy and meteorology at the University of Graz, and to apply his nimble mind to any number of varied interests, among them his continuing romance with Greenland. In 1930 he again set out for his beloved wasteland, this time as leader of a group of 21 scientists and technicians. With the backing of the Emergency Association of the German Sciences (so named because it was struggling to fund scientific research despite Germany's economic collapse), they were to spend 18 months on the icecap gathering climatological, glaciological and geophysical information.

They planned to set up three camps: one on the western boundary of the icecap; one on the eastern edge; and a third at a station to be called Eismitte, or "middle of the ice," 250 miles inland at lat. 71° N. No expedition had ever attempted to winter that far north or that far inland on the ice sheet. The team arrived in Greenland in April of 1930, and while Wegener saw to the setting up of the westernmost camp and directed proceedings from there, small detachments went off to set up the other two camps. Among those heading for the isolated site of Eismitte were his former student Johann Georgi and a glaciologist, Ernst Sorge. Travel conditions proved to be so brutal that a number of supply runs to Eismitte—including

the one that would have provided the camp with a radio transmitter—had
to be canceled.

Two months after Eismitte had been established, two expedition members returned to Wegener's camp from a supply run to the inland camp to
report that Georgi and Sorge were in dire need of food and fuel, without
which they would be unable to survive the winter. Wegener, whose leadership in the field inspired the same profound loyalty among his teammates
that it did among his students, was well aware that the situation at Eismitte
was serious, and he had already mounted a rescue expedition. With 13
Greenlander guides, 15 dog sledges and a colleague, Fritz Loewe, he set out
for Eismitte on September 21.

The 250-mile journey was a horror of drifting snow and biting wind that
would have tested the endurance of the hardiest youth. During the first
seven days the men and their dogs covered only 38 miles. One after another
the Greenlanders began giving up and returning to the base camp; finally
only one, a tough young fellow named Rasmus Villumsen, remained with
Wegener and Loewe. It took the remaining threesome until the morning of
October 30 to complete the journey to the ice cave that Sorge and Georgi
had hacked out to protect themselves from the awful winter at Eismitte.
Despite the grueling journey, during which Loewe had suffered frostbite so
severe that his colleagues later had to amputate all his toes, Wegener arrived "as fresh, happy and fit as if he had just been for a walk," Sorge
remembered. "Wegener kept exclaiming 'You *are* comfortable here! You
are comfortable here!' over and over again. His energy had not been exhausted by the 40 days of hard sledging; on the contrary, he was fired with
enthusiasm and was ready to tackle anything."

Wegener stayed at Eismitte for two days, eagerly gathering and recording meteorological data. On the morning of November 1, his 50th birthday, the men held a cheerful party in their ice cave, celebrating the occasion
with an apple apiece—rare relief from a diet otherwise made up of canned
and dried food. After the party was over, Wegener and Villumsen set out
for the return to the west camp, leaving the unfortunate Loewe with Georgi
and Sorge to recover during the winter at Eismitte.

His friends never saw the great scientist alive again. Those who had remained at the west camp assumed he had decided to winter at Eismitte, but when he had not returned by April, they sent out a party to make sure he was there. Halfway to their destination they came across Wegener's skis, standing upright in the snow about three yards apart, a broken ski pole between them. Intrigued and not a little concerned, they dug around in the snow but found only an empty supply crate. When they arrived at Eismitte and learned the truth, they quickly returned to the abandoned skis.

Burrowing frantically into the snow and ice, they soon found Wegener's body. Fully dressed, it was lying on a reindeer skin and a sleeping bag, neatly stitched into two sleeping bag covers and covered by another reindeer hide. "His eyes were open," one of the search party reported, "and the expression on his face was calm and peaceful, almost smiling." It did not appear he had died from either hunger or exposure, and his friends concluded that the most likely cause of death was a heart attack, probably induced by the exertion of the journey. Villumsen, the trusty Greenlander who had accompanied him, had obviously buried Wegener with great care and marked the grave—only to disappear himself somewhere in the icy wilderness. Wegener's sorrowing companions placed his body back in the snow exactly as they had found it. They draped black flags on the upended skis, erected a cairn of ice blocks and fashioned a small cross from the broken ski pole. Thus adorned, the barren ice of Greenland became—with singular appropriateness—Alfred Wegener's last resting place.

The obituaries were laudatory and filled with lavish praise for Wegener's achievements as a meteorologist and explorer. Much was written about his expeditions to Greenland, his distinguished career as a scientist and teacher, his abilities as a leader and his academic brilliance. Hardly a mention was made of continental drift, which by then was generally considered to be nothing more than a bizarre fantasy—an aberration in an otherwise exemplary life. Ω

On November 14, 1963, at 7:15 a.m.,
crewmen aboard a fishing vessel 20
miles off the south coast of Iceland were
surprised to see black smoke billowing
from the sea. Assuming that a volcano
was erupting far below on the ocean
floor, they summoned geologists to the
scene. That evening, observers spot-
ted a ridge of hardening lava just beneath the waves, and by the following
morning, a tiny island had emerged
above the surface.

The new island was christened Surtsey—after Surtur, the Icelandic god of
fire—and it continued to erupt off and
on for three and a half years, eventually attaining an area of approximately one square mile. Like Iceland to
its north, Surtsey provided scientists
with one of the rare surface manifestations of the Mid-Atlantic Ridge, the
12,000-mile-long submarine mountain
range that stretches the length of the
Atlantic Ocean.

Surtsey and Iceland not only are parts
of the Mid-Atlantic Ridge, but owe
their very existence to the molten rock,
or magma, that wells up through the
rifts along the ridge. At least 20 million years ago, scientists believe, Iceland
rose from the sea floor much as Surtsey
did in 1963—albeit on a far grander
scale. Continuous spreading—accompanied by eruptions—along Iceland's
section of the ridge widens the country
by about one inch each year.

Iceland's singular landforms provide
dramatic evidence of its fiery birth and
restless nature. One third of the country's 40,000 square miles is volcanically active and is dominated by numerous craters and lava fields. Elsewhere
magma too far below the surface to
create volcanoes heats the rock above,
sending groundwater percolating to the
surface in hundreds of hot springs. Iceland's so-called "fire beneath" also has
pronounced effects on the country's gla-
ciers: It can melt a glacier slowly, safely
feeding rivers and streams, or it can dis-
solve a glacier quickly, causing catastrophic floods.

Molten lava and steam spew from Iceland's
island of Surtsey during a major eruption on
August 27, 1966. The island's dramatic
genesis in 1963 gave scientists their first good
look at a new landform rising from the sea.

A landscape 20 miles southwest of Reykjavik
exhibits two kinds of land-building activity.
The volcano in the background was formed
under a glacier during the last ice age; the lava
field in the foreground, now overgrown
with moss, was created after the ice had
receded, when magma escaped to the surface
through a long rift, or fissure, in the earth.

The now-tranquil Laki Craters of southern
Iceland mark the site of a series of volcanic
eruptions that devastated the island between
June 1783 and February 1784. Gases from
the volcano poisoned Iceland's pasturelands and
killed two thirds of the livestock, triggering
a famine that claimed one fifth of the population

Lava from a 1973 eruption cools and hardens
into crust on the island of Heimaey, off Iceland's
southern coast. As gases charged with sulfur
compounds escaped through fissures into the cool
air, some of the compounds were deposited
on the lava, staining it bright yellow.

The landscape of southern Iceland evinces
the island's contrasting worlds of fire and ice
as the hot springs of an active geothermal
area spout steam amid fields of melting snow.

Streams of meltwater etch twisted courses
downhill from a glacier in southern Iceland.
The region's underground heat speeds
glacial melting and sometimes provokes
sudden floods known as glacier bursts.

Chapter 3

ANSWERS FROM THE ABYSS

On a snowy November afternoon in 1934, two men wearing derbies and fur-collared coats made their way across the Bethlehem, Pennsylvania, campus of Lehigh University. One of the men, Richard Field, was a distinguished associate professor of geology at Princeton University; the other, William Bowie, was an official of the U.S. Coast and Geodetic Survey. Their destination: the cluttered basement office of a 28-year-old physics instructor named Maurice Ewing.

A trim Texan with a shock of unruly hair and owlish spectacles, Ewing had earned his doctorate in 1931 at Rice Institute—where he played first trombone in the school marching band—with a thesis on the paths that seismic waves took through the earth. During the years that followed, he had continued his studies in the young science of seismology by analyzing the complex seismograph tracings produced by explosions set off in the limestone deposits of the Lehigh Valley, and even in the ice of a frozen pond. By noting the behavior of the different kinds of seismic waves produced, by comparing their velocities and reconstructing how they had been refracted and reflected by subterranean layers of materials, Ewing had become an expert in explosion seismology. Field and Bowie knew of his research from papers the young scientist had presented at meetings of the American Geophysical Union. Now they had come to Bethlehem to suggest that he focus his work on solving a particular mystery. Ewing listened intently to his visitors' proposal.

They were interested in the nature of the continental shelf that extends seaward from every land mass, then drops off steeply into the abyss. Was the shelf edge a permanent geological feature, the true edge of the continent, or was it simply the farthest reach of sediment laid down from the land? When Field and Bowie asked Ewing whether he could apply explosion seismology to determine the composition of the shelf, Ewing immediately declared that he could and would. Later he would observe: "If they had asked me to put seismographs on the moon instead of the bottom of the ocean I would have agreed, I was so desperate for a chance to do research."

With the encouragement of his two mentors, Ewing obtained a $2,000 grant from the Geological Society of America, recruited two colleagues to assist him, and borrowed a sailing ship, the *Atlantis,* from the Woods Hole Oceanographic Institution in Massachusetts. For two weeks in October of 1935, he made a series of seismic measurements of the continental shelf off Norfolk, Virginia, by lowering recording devices onto the sea floor, then

In 30 feet of water off the eastern coast of Hawaii, a diver watches molten rock from a volcanic eruption harden into bulbous shapes known as pillow lava. The discovery of similar formations near huge rifts in the deep ocean floor in the 1960s helped to substantiate the concept of sea-floor spreading.

dropping TNT blasting charges overboard from a whaleboat and detonating them at depths of about 600 feet.

Ewing's analysis of the seismograms indicated that the continental shelf was not a permanent feature but was composed of sedimentary deposits piled as high as 12,000 feet above the sloping bedrock, or basement, of the ocean floor. Neither he nor his sponsors knew quite what to do with these findings. It seemed likely that oil deposits lay in the offshore sediments, but when Ewing sought oil-company support for further studies, he was informed that oil was so readily accessible on land that there was no point in looking for it under the sea.

The expedition nonetheless kindled in Ewing a lifelong passion for plumbing the sea bottom. In the coming decades, he and a growing contingent of loyal aides would devote nearly every waking hour to the exploration and analysis of undersea regions. Inspired in no small measure by his indefatigable example, others, too, probed the bottoms of the oceans. With the aid of ever more sophisticated equipment, oceanographers and seismologists accumulated findings that astounded scientists around the world and reopened the languishing debate about continental drift.

Although water covers approximately 70 per cent of the earth's surface, the deep ocean floors were virtually unknown territory when Ewing began his oceanic investigations. Nineteenth Century oceanographers, leaning over the rails of their ships and dropping weighted lines to take soundings of the bottom, had only the vaguest idea of what lay beneath their feet. Even the first echo sounders of the 1920s revealed little of significance about the sea floor, because they were notoriously unreliable and their range was limited.

Maurice Ewing's explosion-seismology techniques promised a bounty of new information, but larger events forestalled the payoff: The outbreak of World War II brought all civilian ocean research to a standstill. At the same time, though, the global conflict spurred the development of new technology designed to probe the undersea realm. Sophisticated sonar equipment was developed to search for the enemy U-boats that were preying on Atlantic shipping convoys, and depth recorders helped locate suitable landing sites for amphibious assaults on hostile shores.

After the War, such equipment was in greater demand than ever. Because the United States government believed that submarines would play a major role in any future conflict, federal funds were poured into a program to map the world's ocean floors. Precision depth recorders, accurate to within a few feet and capable of laying out continuous profiles of the sea floor, were eventually fitted to dozens of oceanographic research ships and naval vessels.

Ewing was in the forefront of these developments. In late 1940 he left his post at Lehigh and went to the Woods Hole Oceanographic Institution, where he worked throughout the War on underwater acoustics projects for the Navy. With the coming of peace, he moved with some of his Woods Hole associates to Columbia University and, financed by the Navy, resumed his oceanographic surveys. Then, in 1947, the National Geographic Society commissioned him to explore the Mid-Atlantic Ridge and the sea floor around it. Little was known about this undersea region, and Ewing equipped himself not only with explosion-seismology and depth-

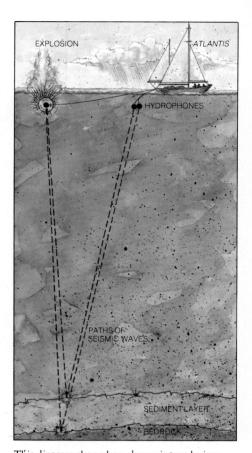

This diagram shows how dynamite explosions were used to measure ocean-floor sediments in the 1940s and 1950s. Seismic waves reflected from the sediment layer reached hydrophones trailed astern of the ship before the waves reflected by the bedrock. The time difference indicated the thickness of the sediment.

sounding gear, but with devices for sampling sea water, monitoring ocean temperatures at various depths and bringing up cores of sediments from the sea floor itself.

At the time, most earth scientists—Ewing included—believed that ocean sediment would offer fossil clues to the entire history of the earth. But the first core samples, taken from almost a mile down, were perplexing. They contained a layer of recent sediment lying directly on another layer that was more than 20 million years old; inexplicably, there was no trace of material from the intervening millennia.

Seismic tests produced more surprises. Using hydrophones deployed on the sea floor to pick up seismic waves reflected and refracted from the deep ocean floor, Ewing discovered that the sediment was much thinner than expected. Some scientists had predicted that eroded soil and dead ocean organisms accumulating for some three billion years would have built up a layer of sediment at the bottom of the sea as much as 12 miles thick. But

Hunched over his communications and recording gear, geologist Maurice Ewing supervises a seismic "shoot" aboard the research ship *Atlantis* in 1951. Ewing's discovery that sea-floor sediments were surprisingly thin provided one of the first indications that the ocean floors are constantly being renewed.

An underwater explosion sends up a plume of spray near the *Atlantis* as a crewman pays out the line towing the hydrophones. To record the seismic waves effectively, the instruments had to be kept motionless in the water.

beyond the continental shelves, the thickest layers Ewing could find measured only a few thousand feet, representing sedimentation during a mere 100 million to 200 million years.

When he reached the Mid-Atlantic Ridge, Ewing sought to take core samples but was thwarted when his coring equipment was damaged in heavy seas. He then lowered a dredge and dragged it across the craggy slopes for four hours. Seismic tests had indicated to Ewing that the area of slopes nearest the ridge center had only a thin covering of sediment, but he was still somewhat startled by what he found when he hauled up the dredge. Instead of sediment, it bore a cargo of rocks that had apparently been subjected to severe heat or pressure; a second dredging turned up shiny globules of pillow lava, which is formed when molten rock is quenched in water. It appeared that the ocean floor was much younger than anyone had imagined; not only was it thinly sedimented, but its rocks seemed newly formed—and of volcanic origin.

Ewing launched two more expeditions to the Mid-Atlantic Ridge in 1948, and in the following year he founded Columbia's Lamont Geological Observatory in a Hudson River mansion donated to the university by the widow of banker Thomas W. Lamont. There, Ewing gathered about him a group of graduate students and other researchers who assisted him in his studies of the ocean floor. And the surprises kept on coming: Scientists knew little about the bedrock of the sea floor, but many geologists assumed that it was composed of sunken continental material. On a cruise from Woods Hole to Bermuda not long after Ewing had set up the new observatory, Lamont staffers came up with seismic data indicating that the oceanic crust is composed of dense basalt and is only three miles thick. (The largely granitic crust underlying the continents is some 25 miles thick.)

In 1953, frustrated by the need to beg and borrow ships for short periods of time from other institutions, Ewing acquired a 202-foot vessel, the *Vema*, for the sole use of the observatory. Once a luxury yacht belonging to the cereal-company heiress Marjorie Merriweather Post, the craft had been taken over by the U.S. Navy during World War II and had since fallen upon bad times; at one point, it was stripped and used as a floating dormitory. Under Ewing's direction, it was refitted for duty as an ocean research vessel. In his eagerness to gather great volumes of data, Ewing kept the *Vema* at sea as much as possible, bringing up seabed samples, taking depth soundings and seismic readings.

In the meantime, Ewing had decided that it was time to translate the available echo-sounding profiles of the North Atlantic sea bottom—six widely separated studies in all—into a topographical map of the ocean floor. He assigned the job to Bruce Heezen, a graduate student who had studied geology at the University of Iowa; Heezen, in turn, enlisted the aid of a cartographer at the observatory, Marie Tharp. "As the preliminary sketch emerged," Heezen later wrote, "Miss Tharp was startled to see that she had drawn a deep canyon down the center of the Mid-Atlantic Ridge."

For a time, Heezen doubted the existence of such a large-scale rift. Then in 1953, while studying the potential for earthquake damage to transatlantic telephone cables, he asked another staffer in the Lamont drafting department to plot the location of all recent Atlantic earthquakes; for convenience, he suggested using as a base map the chart that Marie Tharp was preparing. As the dots went onto the map, Heezen noticed that they fell within the rift valley that Tharp had drawn in the ridge.

When the coincidence was brought to Ewing's attention, he began to think that his young associates were onto something; he and Heezen began gathering and plotting data on midocean earthquakes from every source they could find. In time, their chart showed a band of earthquakes running through the middle not only of the ridge and rift valley mapped by Tharp, but of all the world's oceans. The connection between underwater earthquakes and the previously mapped rift was so convincing that the two scientists predicted, from seismic activity, the existence in some oceans of rifts and ridges that had not yet been discovered.

In 1956, Ewing and Heezen disclosed some of their conclusions about a global rift in a paper presented to the American Geophysical Union. Many of their fellow scientists at first rejected the idea; it implied, after all, a widening of the earth, rather than the generally accepted contraction. But the notion was largely confirmed when it was put together with some altogether different studies already done by the noted British geophysicist Sir Edward Bullard.

Geophysicists had long known that radioactive materials in the earth generate heat, and they assumed that continental crust generates more heat than does the ocean floor (the granite of the continents contains a much greater abundance of radioactive elements than does the basalt of the oceanic crust). It seemed reasonable, therefore, to expect that more heat would be radiated from the continents than from the sea floors. But Bullard, using a special probe with thermometers mounted along its length, found in the early 1950s that the average heat flow from the ocean bed is approximately the same as from land. Soon afterward, he took a series of measurements along the Mid-Atlantic Ridge and discovered that the rate of heat flow along the crest of the ridge—where Ewing and Heezen later claimed there was a valley—was as much as eight times greater than elsewhere on the ocean floor. The obvious explanation was that the ridge must lie atop a great crack in the earth's crust where new, hot material was rising to the surface.

Ewing, Heezen and Tharp published their description of the North Atlantic sea floor in 1959. The U.S. Navy had sponsored much of their work and, for reasons of national security, would not permit the disclosure of precise depth soundings that might be of use to a potential enemy. So the accompanying map was not a conventional contour map but a physiographic diagram, a sort of bird's-eye view drawn in three dimensions *(pages 78-79)*. The map showed, for the first time, a breathtaking panorama of vast plains, peppered here and there with conical features known as seamounts, that swept toward an enormous mountain range that was wider and more continuous than any mountains above the sea. This was the Mid-Atlantic Ridge, and through it, 12 miles wide in some places, ran the great and still-mysterious valley.

As more and more of the world's sea floors were explored during the ensuing few years, the Lamont researchers used the accumulated data to expand their map until it showed beyond a doubt that the ocean ridges were not isolated features, but part of a 40,000-mile-long mountain range that curls through all the oceans of the world, encircling the planet like the seam of a baseball. The length, width and height of this Mid-Ocean

Ridge, as it was soon named, made it the predominant topographical feature of the earth.

Comprising millions of square miles of jagged peaks, jumbled cliffs and valleys, the ridge snakes in a great S bend down the center of the Atlantic, almost reaches the Antarctic, then curves east around the African continent and into the Indian Ocean, where it forks in two. One chain goes north into the Gulf of Aden, where it splits again into the Red Sea and the East African rift system; the other strikes off south of Australia and New Zealand into the Pacific where, several thousand miles from the tip of South America, it curves north, eventually entering the Gulf of California and joining the San Andreas Fault zone. Almost entirely hidden under the ruffled surface of the oceans, its great peaks thrust a mile or more above the rift valley floor, occasionally poking above the waves to form rocky islands like the Azores and Iceland.

"The discovery that numerous, previously known, individual ridge systems were all part of the same worldwide system is probably the most exciting major discovery about earth science in the past 20 years," said Harris B. Stewart of the U.S. Coast and Geodetic Survey in 1959. Certainly, those scientists who had dismissed the jigsaw-puzzle fit of South America and Africa as mere coincidence now found it difficult to explain why the curving ridge lay precisely halfway between, and parallel to, the coasts of the two continents. At the very least, the discovery called into question traditional theories about the earth, none of which had predicted, or could explain, the ridge's existence. For the time being, however, the role of the Mid-Ocean Ridge remained a matter of speculation; all that was known for sure was that a giant crack had been found in the crust of the earth.

During the Second World War, Princeton professor Harry H. Hess had commanded an attack transport, the U.S.S. *Cape Johnson,* in the Pacific. In preparation for the landings at Iwo Jima, his ship was fitted with a new device called a fathometer—an echo sounder that drew continuous profiles of the sea bottom and enabled troop-carrying ships to move as close inshore as possible. Hess was so delighted with it that he kept it switched on day and night for the whole time he was at sea. He was, after all, a geologist and not the kind of man to pass up this unprecedented opportunity to explore the floor of the Pacific.

As he studied the profiles for hour after hour, Hess noticed a number of strangely shaped submarine mountains scattered about the western Pacific. He named them guyots, after the distinguished 19th Century Princeton geologist Arnold Guyot, although at first he did not know what to make of them. Like most old volcanic islands subjected to eons of scouring by wind and waves, the guyots had extremely flat tops. What was so puzzling about the guyots was that they were all well below the surface of the ocean, far beyond the reach of erosion.

Hess located about 20 guyots with the *Cape Johnson's* fathometer, but hundreds more were discovered after the War by oceanographic research vessels. And their shape was not the only mystery; core sampling and depth sounding showed that the farther the guyots were from the Mid-Ocean Ridge, the deeper they were. The only explanation, Hess concluded, was that they had formed as volcanic islands along the ridge, had been truncated by erosion while on the surface but then, somehow, had submerged and drifted away from the ridge.

Hess, like most of his academic colleagues, placed no credence in the theory of continental drift, so he did not imagine for a moment that guyots moved as a result of some global waltz. But the suggestion by Maurice Ewing and Bruce Heezen in 1956 that the midocean ridges were part of a continuous underwater mountain chain set Hess to thinking once more about the origin of guyots and their role in what might be happening under the sea. Indeed, the rapid confirmation that the planet was girdled by a continuous submarine mountain range prompted scientists everywhere to try to make some sense out of this awesome feature. One popular theory, subscribed to by Heezen among others, was that the earth was somehow expanding and that the rift was equivalent to the cracks in the shell of a boiled egg.

Hess cautiously rejected the idea of an expanding earth for lack of supporting evidence, and looked instead for an explanation that would account not only for his guyots, but for all the other perplexing discoveries of modern oceanography—the absence of ocean-bottom rocks older than about 150 million years, the missing sediment, the heat flow from the rift and the thinness of the oceanic crust. After wrestling with the problem for many months, he produced a startling and original hypothesis: that the ocean floors were moving like conveyor belts, carrying the continents along with them.

Mindful of the conservatism of the scientific establishment, and no doubt recalling what had happened to Alfred Wegener in a similar situation, Hess carefully hedged his radical ideas when it came time to commit them to print. His paper, written in 1960 and widely circulated among scientists before its publication in 1962, was sedately titled "History of Ocean Basins." Hess began with a disarming disclaimer: "The birth of the oceans is a

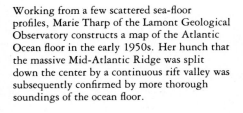

Working from a few scattered sea-floor profiles, Marie Tharp of the Lamont Geological Observatory constructs a map of the Atlantic Ocean floor in the early 1950s. Her hunch that the massive Mid-Atlantic Ridge was split down the center by a continuous rift valley was subsequently confirmed by more thorough soundings of the ocean floor.

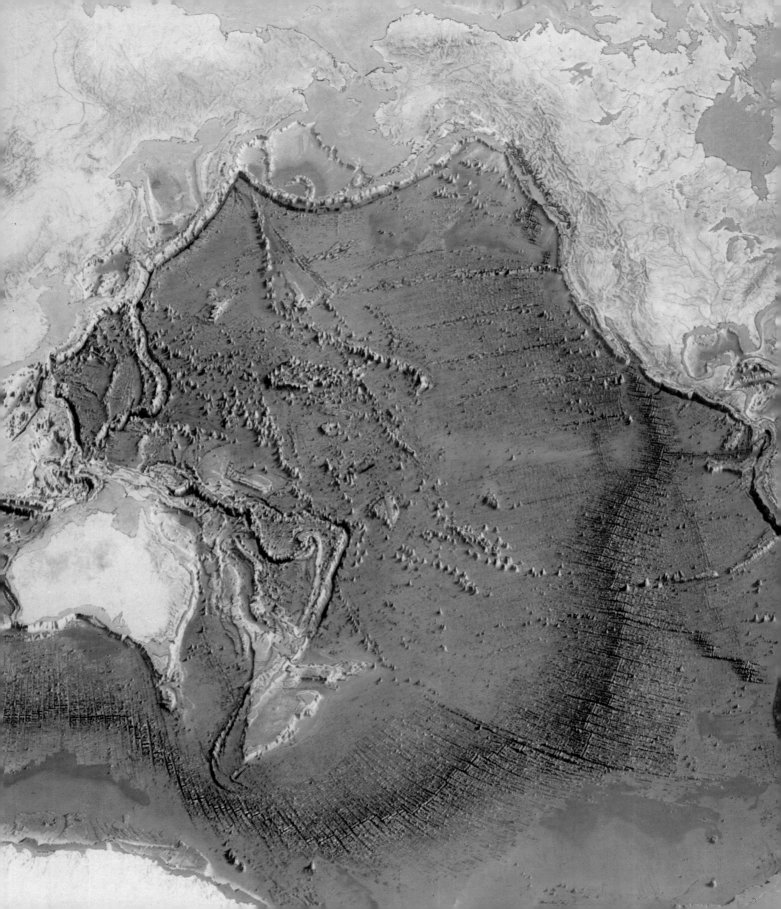

This stunning map produced by Marie Tharp and Bruce Heezen in 1977 was the first to depict the awesome scale of the Mid-Ocean Ridge system, whose 40,000-mile length dwarfs that of any mountain range found on land.

matter of conjecture, the subsequent history is obscure, and the present structure is just beginning to be understood." And then he warned his readers that what he was about to present was "an essay in geopoetry" that bordered on fantasy.

Hess's fundamental proposition was that the sea floor is not permanent, but is constantly being renewed. The Mid-Ocean Ridge is indeed a crack in the crust, he said; through it, hot material from the underlying mantle continually wells up and spreads outward. (Commenting on the theory a year after Hess had written his paper, another geologist, Robert Dietz, would coin the term "sea-floor spreading.") Hess estimated that new crust is generated at the rate of about half an inch per year on each side of the ridge. At this pace, all the ocean floors of the world would have been formed during the last 200 million years—less than 5 per cent of the earth's geologic history.

Obviously, a spreading sea floor would constantly increase the size of the earth unless something else was going on, and since there was no evidence to indicate that the world was getting bigger, Hess suggested that as new ocean crust is created along the ridge, old crust is simultaneously destroyed in the deep ocean trenches that lie near the edges of continents. He theorized that some powerful force is pulling the old ocean floor down into the earth's mantle at the trenches, a process later known as subduction.

The existence of such a force—and an explanation for it—had been suggested in the 1930s by the Dutch researcher Felix A. Vening Meinesz. Scientists had long assumed that the trenches were held down in the mantle by the weight of a dense, underlying material, but Meinesz, working in deep-diving submarines, had conducted gravity tests that indicated that there was no such material present. The Dutch scientist decided that his findings could be explained by the circulation of heat in the earth's interior. He theorized that convection currents in the mantle, heated and rising at the ocean ridges, cooling and descending at the trenches, must be responsible for pulling the ocean floor down into the abyssal trenches. Hess took the idea one step further and proposed that such currents provided the driving force for sea-floor spreading.

Hess's theory would account for the relatively thin sediment layer on the ocean floor, and for the absence of rocks of more ancient vintage. "The whole ocean is virtually swept clean every 300 to 400 million years," Hess wrote. The process also accounted for the mysterious guyots: Rising as conical volcanic islands along the ocean ridges, they had been flattened by erosion before being transported by the moving sea floor away from the ridges and into deeper water.

Hess brought the various strands of evidence together into a radically different vision of the earth's major features: "The ocean basins are impermanent features, and the continents are permanent, although they may be torn apart or welded together and their margins deformed. The continents are carried passively on the mantle with convection and do not plow through oceanic crust."

Hess's "History of Ocean Basins" was widely read by scientists and students. Many were intrigued by his originality, but most considered his ideas farfetched, to put it kindly. For more than 100 years, scientists had clung to the doctrine of permanent ocean basins and had scoffed at the notion of drifting continents. They could not be swayed to the opposite

The Fruitful Failure of "CUSS 1"

In March of 1961, on the eve of the first manned flight into space, a converted Navy barge was towed out of San Diego harbor to explore a less distant but no less elusive frontier—the interior of the earth. Christened *CUSS 1*, the vessel hardly looked like a scientific marvel; author John Steinbeck, on board to report on its mission, credited it with the "sleek race lines of an outhouse standing on a garbage scow."

The unprepossessing ship was embarking on the initial phase of a plan to drill right through ocean crust and into the earth's mantle. Scientists had dubbed the project Mohole, after Andrija Mohorovičić, who first deduced the location and density of the mantle's top from the behavior of seismic waves.

The effort to pierce the mantle had to be made below the deep-sea floor, where the crust is much thinner than under the continents. *CUSS 1's* assignment was to test the technology of deep-sea drilling while collecting core samples that could throw new light on the composition and movements of the crust.

When the barge had been positioned near Guadalupe Island, the crew lowered a string of drill pipes to the ocean floor 12,000 feet below and began working down through the soft sediment. At a depth of more than 550 feet, the diamond-studded drill bit met stiffer resistance, and a core sample—including a blue crystalline chunk of basalt from an ancient lava flow—was brought up.

Having shown that it could obtain rock specimens through two miles of water, *CUSS 1* returned to port, its mission accomplished. But its trial made it clear that difficulties lay ahead. Drilling deeper would require both a larger vessel to carry additional pipe and expensive new technology to guide the drill. Some scientists argued that the money would be better spent on less problematic objectives, and in 1966 Congress cut off funding for Mohole. Still, by proving that drilling the deep-ocean floor was possible, the voyage of *CUSS 1* held out the first real hope for direct observation of what lay beneath the earth's crust.

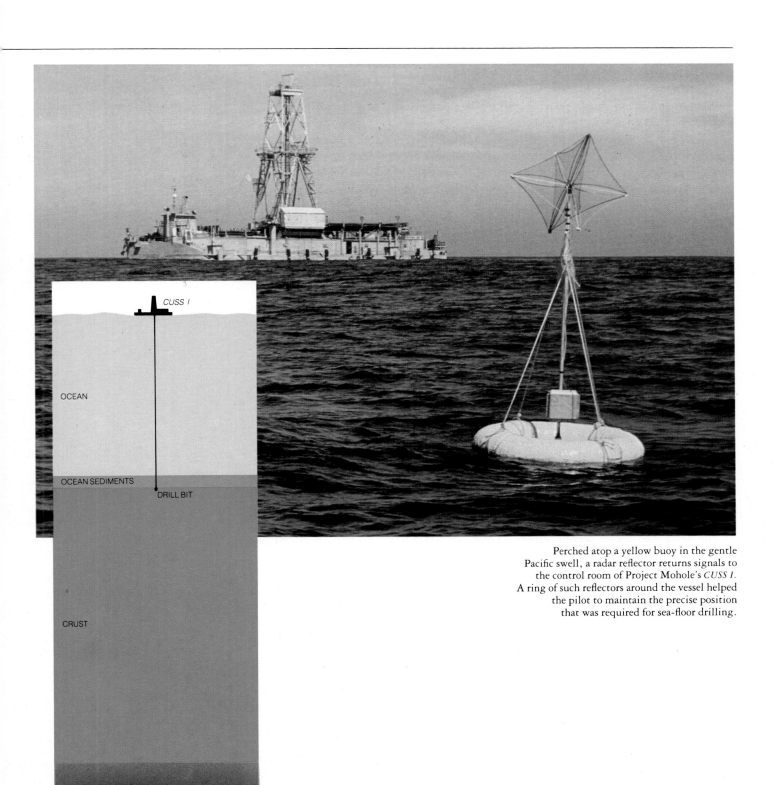

OCEAN

OCEAN SEDIMENTS

CRUST

MANTLE

CUSS I

DRILL BIT

Perched atop a yellow buoy in the gentle
Pacific swell, a radar reflector returns signals to
the control room of Project Mohole's *CUSS I.*
A ring of such reflectors around the vessel helped
the pilot to maintain the precise position
that was required for sea-floor drilling.

A chart traces what *CUSS I* accomplished off
Guadalupe Island—and what remained to be
done. The crew managed to guide the drill's
bit through more than two miles of water and
sediments to a layer of dense basalt. But
when drilling halted, the elusive mantle was
still more than five miles away.

point of view by passages of "geopoetry"; their minds would be changed only by hard, scientific evidence.

Even a geopoet could not have foreseen that, all along, the required evidence had been registered, as if on an enormous tape recorder, by the earth itself.

Scientists do not know for certain what produces the earth's magnetic field—the lines of magnetic force that run between the North and South Poles. The most widely accepted theory is that the field is somehow generated by the circulation of molten iron in the core of the earth. It is, of course, not necessary to understand a phenomenon in order to make use of it; even early mariners knew that their magnetic compasses would always point to the north. They understood that both magnets and the earth itself have north and south poles, and that unlike poles attract each other while like poles repel each other. Thus the south, or north-seeking, pole of a compass needle is always drawn toward the North Magnetic Pole. At the same time, the north-seeking end will dip toward the ground in the Northern Hemisphere, while the south-seeking end will dip in the Southern Hemisphere. In either case, the angle at which the needle slants—called inclination—increases with proximity to the poles.

Some rocks, too, have compass-like properties; as molten lava solidifies and cools, for example, tiny crystals of magnetite within it become permanently magnetized. Achilles Delesse, a French physicist, was the first to observe—in 1849—that such rocks were magnetized in parallel with the earth's magnetic field, as if the rocks were recording compasses. This knowledge raised the possibility that magnetic rocks could be used to trace the small changes that were known to occur over time in the positions of the magnetic poles: If the ages of the rocks were known, scientists could determine by the direction of their magnetism the location of the magnetic poles at the time the rocks were formed. Then, in 1906, the French physicist Bernard Brunhes made the startling discovery that some rocks are magnetically oriented in exact opposition to the earth's field. Brunhes proposed that this strange phenomenon was the result of a past reversal in the polarity of the global magnetic field.

Geologists in many parts of the world continued to survey the magnetism of rocks, but the new science of paleomagnetism—the study of the history of the earth's magnetic field—did not really come into its own until the early 1950s, after the British physicist Patrick M. S. Blackett invented a sensitive device called the astatic magnetometer. Using this apparatus, it was possible for the first time to detect the orientation of extremely weak magnetic fields, enabling researchers to conduct paleomagnetic studies of types of rocks whose magnetism could not be discerned by earlier equipment.

Blackett, a physicist at the University of London, had won the Nobel Prize in 1948 for his work in nuclear physics and cosmic radiation. He developed his magnetometer to conduct experiments into the origins of magnetism, but soon turned to paleomagnetics. He had long been interested in the concept of continental drift and believed that magnetic measurements might help determine whether the great land masses had indeed shifted their positions.

Blackett and his colleagues analyzed rocks that had been gathered from all over the British countryside (their original positions in the earth were carefully noted before they were brought to the laboratory). The researchers discovered that the magnetic orientation of many of the rocks, depending on their age, deviated as much as 30 degrees from the modern position of the North Magnetic Pole. More surprising, the rocks' magnetic inclination was much less than it should have been, indicating that the rocks must have been formed in a more southerly latitude. To Blackett and his associates, the conclusion was inescapable: Their paleomagnetic findings could be explained only if Britain had rotated clockwise through 30° of longitude and drifted northward at the same time from a lower latitude, starting some 200 million years ago.

Another research team, led by S. Keith Runcorn at the University of Newcastle-upon-Tyne, thought otherwise. Runcorn, a former radar scientist, doubted that continents had moved; he believed instead that the earth as a whole had shifted its position relative to the magnetic axis, causing the poles to "wander." By plotting the position of the magnetic poles going back through time, as indicated by the paleomagnetism of rocks in Britain and on the Continent, Runcorn's team traced the wanderings of the North Magnetic Pole along a 13,000-mile path that started in western North America about one billion years ago, curved across the northern Pacific and northern Asia, and finally reached its present position in the Arctic about 20 million years ago.

Runcorn presented his findings in 1955, commenting that they did not indicate continental drift. He was soon to change his mind. Similar measurements, taken in North America, indicated that the North Magnetic Pole had during the same period of time followed a similar course, but in a completely different location: 30° of longitude separated it from the path charted in Europe. Runcorn realized that the only way to explain such a divergence was to suppose that when the North American rocks had formed, they had been 30° of longitude closer to the European rocks—that there had been no Atlantic Ocean, and that the two continents had been joined. When this was assumed, the plots of the separate paths matched almost perfectly. The poles may have wandered, he concluded, but the continents had also drifted.

Further paleomagnetic comparisons elsewhere around the world strongly indicated that all the southern continents except Antarctica had drifted steadily northward, and thus lent new credence to the old idea of Gondwanaland as the mother of all southern continents. In India, the evidence was particularly impressive. Paleomagnetic analysis of successive layers of lava on the sprawling Deccan Plateau presented a timetable of the subcontinent's journey from the Southern Hemisphere: The oldest layer, some 150 million years old, contained magnetic evidence that placed India below the Equator at that time; each subsequent layer recorded a slightly smaller inclination as the great land mass moved steadily north.

The scientific establishment was not particularly impressed by these findings, and for good reason—the science of paleomagnetism was and remains an inexact one. Rocks are at best undependable recorders of the magnetic field, and interpreting their secrets requires numerous tests with plenty of room for error. Many scientists thought that the paleomagnetic evidence for continental drift was based on inadequate sampling, inaccurate measurements and unjustified assumptions.

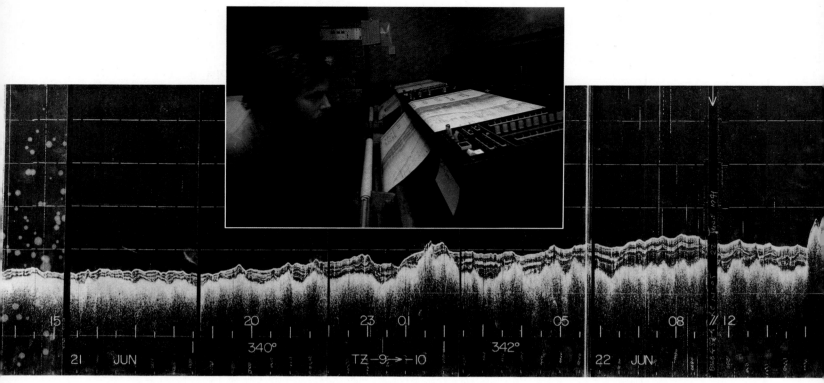

A continuous profile of the ocean floor north of Japan shows the 29,000-foot-deep Kurile Trench *(right)*, where the Pacific Plate *(above)*, its layers of sediment clearly visible, is diving beneath the Eurasian Plate. Such profiles are made by filming—on equipment like that shown in operation aboard the *Glomar Challenger* in 1980 *(inset)*—a continuous record of seismic waves reflected from the sea-floor material.

It was especially easy to distrust the paleomagnetic arguments when they led to such improbable speculations as the possibility that the earth's magnetic field has in the past reversed itself, transposing the north and south poles. Yet in the early 1960s, even that proposition gained respectability when Allan Cox, Richard Doell and Brent Dalrymple, scientists at the U.S. Geological Survey, collected and studied tons of rocks from all over the world. Using the latest radioactive testing methods to date their samples, Cox, Doell and Dalrymple identified nine magnetic reversals during the last three million years. Later paleomagnetic studies would indicate that the earth has switched its magnetic polarity at least 171 times in the past 76 million years. While no one knows for certain what causes these reversals, they are most likely the result of disturbances in the earth's molten core.

Such reversals take place over periods of 5,000 to 10,000 years, while the polarity, once established, can last a million years or more. The most recent recorded magnetic reversal ended some 700,000 years ago. Whether another one will begin in the near future (geologically speaking) is unknown: These events have occurred so irregularly that it is not possible to discern a pattern for prediction. In any case, a reversal would pose few problems for navigators; they could readily adjust their compass readings to compensate for the changes taking place during the long reversal process.

The magnetic mysteries were soon compounded by new findings on the ocean floors. In the mid-1950s, a U.S. government research ship had carried out a detailed survey of a 280,000-square-mile patch of the northeast Pacific; as it steamed slowly back and forth across the area on parallel courses five miles apart, it had towed in its wake a "mag-fish"—a torpedo-like instrument used to measure the magnetic intensity of the earth's magnetic field. Arthur D. Raff, from the Scripps Institution of Oceanography in California, and Ronald G. Mason, a British scientist visiting Scripps, spent months collating the data from the magnetometer. Gradually they discerned a curious striped pattern of an alternating strong and weak magnetic field. Stripe was added to stripe until at last the whole chart of the

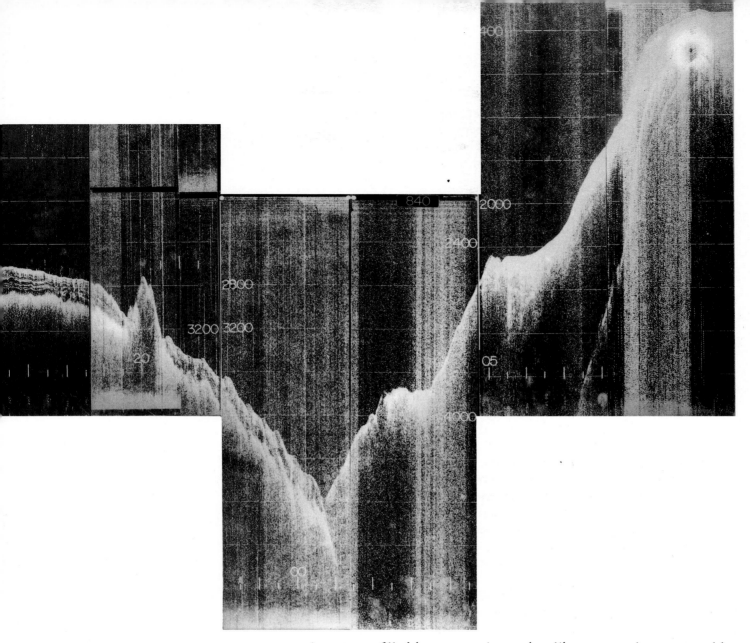

research area was filled by an amazing, zebra-like pattern that ran roughly parallel to the coastline; later surveys disclosed similar magnetic patterns in adjacent areas of the Pacific. No one had ever seen anything like it before, and no one knew what to make of it.

At Cambridge University in England, the presence of Sir Edward Bullard—the discoverer of the high heat flow from the Mid-Atlantic Ridge—had attracted an enthusiastic team of young scientists who were anxious to continue research into the controversial subject of continental drift. Among them were Frederick J. Vine, a graduate student, and his supervisor, Drummond Matthews, who in 1961 had written his doctoral thesis on dredged basalts from the North Atlantic. Both men were firm believers in continental drift, Vine claiming to have been converted at the age of 14 when, like Wegener, he had been struck by the fit between Africa and South America.

In 1962, Vine and Matthews took part in the International Indian Ocean Expedition and, on board H.M.S. *Owen,* carried out a magnetic survey of the Carlsberg Ridge, part of the midocean chain. The survey disclosed evidence of the same puzzling, zebra-like patterns of strong and weak magnetic intensity first recorded in the Pacific and since noted in many of the world's oceans. Back in Cambridge, where the two young scientists shared a room in an old stable, they spent endless hours discussing the perplexing geology of the deep oceans. Harry Hess's geopoetry had made a deep impression on them both, and they analyzed his paper over and over again.

One day, while sitting at the tea table in the old stable, they suddenly realized that the proof of sea-floor spreading was probably staring them in the face. The magnetic stripes, some of them more than 20 miles wide, were not necessarily due to variations of *intensity,* as it had first appeared; they might be due to changes in *direction.* Both men were well aware of the documentation on pole reversals; perhaps normal magnetism produced a high-intensity magnetic field, reversed magnetism a weaker one.

If hot mantle material was welling up in the Mid-Ocean Ridge, they reasoned, it would be magnetized in the direction of the earth's magnetic field as it cooled; if the sea floor was spreading, this band of magnetized rock would be steadily carried away from the ridge; if the poles reversed from time to time, then strips of the sea floor parallel to the ridge would be magnetized in alternate directions; and since the dates of pole reversals had been roughly calculated, then the magnetic pattern on the spreading sea floor would be documenting not only the floor's age but also the rate at which it was spreading.

As so frequently happens in science, someone else independently hit on the same idea at about the same time. Lawrence W. Morley, a Canadian geophysicist, submitted to the American *Journal of Geophysical Research* a paper outlining almost exactly the hypothesis constructed by Frederick Vine and Drummond Matthews. It was curtly rejected with a tart note observing that "such speculation makes interesting talk at cocktail parties, but it is not the sort of thing that ought to be published under serious scientific aegis."

Vine and Matthews had collected more data to support their contentions, and enjoyed a better reception. In September 1963 the respected magazine *Nature* published their short article titled "Magnetic Anomalies over Oceanic Ridges." On the whole, though, their findings failed to excite much interest. Most scientists believed that the magnetic poles were no more likely to reverse than the sea floor was to spread; a combination of the two ideas was almost impossible to take seriously.

One scientist who did take it seriously was the Canadian geophysicist Tuzo Wilson, a courtly professor at the University of Toronto. Like most men in his field, Wilson had been a believer in the theory of a cooling and contracting earth with fixed continents and ocean beds; as recently as 1959, he had dismissed continental drift as an idea "without a cause or a physical theory." But paleomagnetic evidence and the sea-floor-spreading geopoetry of Harry Hess had changed his mind, and by early 1964 Wilson was in London to take part in a continental drift symposium sponsored by Britain's Royal Society.

During this landmark gathering—at which a number of previously skeptical scientists found themselves won over to the notion of drift—Sir Edward Bullard showed a striking map of the Atlantic continents joined together as a single land mass. It had been plotted by a computer programed to piece the continents together along the contours of the continental shelves rather than along the more changeable modern coastlines. Alfred

Sailors aboard the American research vessel *Explorer* in 1960 prepare to lower a torpedo-like magnetometer into the water off the Florida Keys. Towed 500 feet behind the ship, the instrument measured the magnetic field emanating from the rocks of the sea floor.

A chart of magnetometer readings taken in the northeast Pacific shows a zebra-like pattern of variations. Scientists at first were baffled by such irregularities, but realized in 1963 that they indicate ancient reversals in the earth's magnetic field that had been recorded in the sea floor as it spread slowly from the ocean ridges.

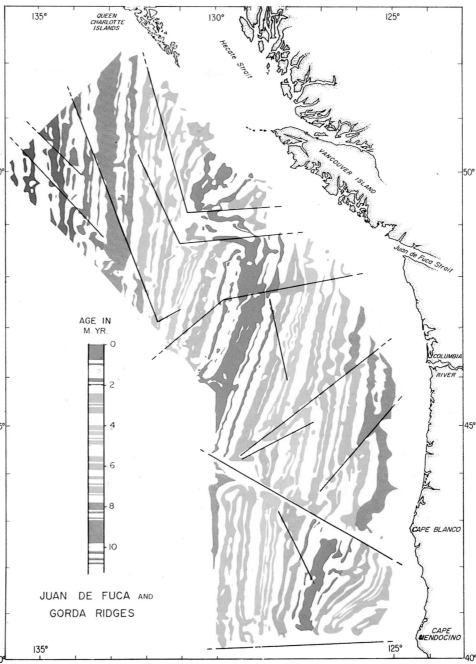

AGE IN M. YR.

0
2
4
6
8
10

JUAN DE FUCA AND GORDA RIDGES

Wegener, too, had proposed such joinings of the continents, but he had lacked the data—and the sophisticated electronic equipment—to make as impressive a case as Bullard presented with his map, which indicated that the continents had once fit together almost as snugly as the pieces of a jigsaw puzzle.

In early 1965, Tuzo Wilson was in England again, this time at Cambridge University—as was Harry Hess, there on sabbatical from his post at Princeton. Naturally enough, the two men got together with Vine and Matthews to discuss some of the geological implications of spreading sea floors. In the course of their conversations, it occurred to Vine and Wilson that the magnetic patterns on either side of a ridge should be symmetrical, since they were formed at the same time and subjected to the same magnetic field. The magnetic survey that Vine and Matthews had taken in the Indian Ocean was not detailed enough for proof of such symmetry, so Wilson and Vine pored over the more precise charts that Ronald Mason and Arthur Raff had worked up earlier for a section of the northeast Pacific. Wilson and Vine studied the magnetic stripes on each side of the Juan de Fuca Ridge, southwest of Vancouver Island, and found that they were indeed remarkably symmetrical, one side being almost a mirror image of the other. Not only that, but Wilson and Vine were able to correlate the widths of the stripes with the time scale of pole reversals during the last four million years.

Scientists at Maurice Ewing's Lamont Geological Observatory had discovered a similar pattern in a section of the Mid-Atlantic Ridge south of Iceland, and in 1965 the National Science Foundation's research ship *Eltanin* revealed an even more striking magnetic profile that showed an almost perfectly symmetrical pattern stretching for more than 1,000 miles on each side of the East Pacific Rise, south of Easter Island. At the April 1966 meeting of the American Geophysical Union, Lamont staffer James R. Heirtzler flashed a slide of this profile onto a screen along with a computer projection of what the profile should look like, given the known sequence of geomagnetic reversals and a supposed spreading rate of nearly two inches a year. So strikingly similar was the comparison that geophysicist Allan Cox, chairman of the conference session at which Heirtzler made his presentation, immediately shed his past doubts about sea-floor spreading. "Good grief!" he thought, "Vine is right after all."

One of the theories that had made sea-floor spreading more tenable was devised by Tuzo Wilson in explaining an odd feature noted in the Mid-Ocean Ridge: Studies of geomagnetic charts revealed that in many places the ridge line is displaced by great fractures, its course sliced into offset segments as it snakes around the globe. Frequently, these fractures—or faults—continue for hundreds of miles on either side of the separated ridge. Most scientists assumed that large-scale geological faults had appeared between adjoining segments of the ocean floor and that the ridge crest, once continuous, had been offset as huge blocks of ocean floor sheared past each other in opposite directions. Called transcurrent faulting, this was a common enough geological phenomenon on land, and there seemed little reason to doubt that it could also occur in midocean.

Wilson, who was always prepared to consider new hypotheses, thought otherwise. If great slabs of the ocean floor had been moved horizontally by hundreds of miles, he asked, what had happened to the displaced ocean floor? There was no trace of it: The active faults seemed to fade into undisturbed sea bottom, which could hardly be the case if enormous segments of crust were sliding against each other all along the length of the fracture. Indeed, seismic data indicated that earthquakes, which are caused by such crustal motions, did not occur along the fault lines extending beyond the ridges; quakes were limited to the fault segments between the offset ridges.

While making a paper model of a fractured ocean ridge and considering how the spreading movement might work, Wilson concluded that the fractures must be a new type of fault, unlike anything else on earth. The offsets were not caused by displacement, he thought, but were original and unchanging features, the result of weaknesses that occurred when the continents first split apart. In a conventional transcurrent fault, the north-south ridge segments would continue to separate, one moving westward, the other moving toward the east; all the while, the adjacent blocks

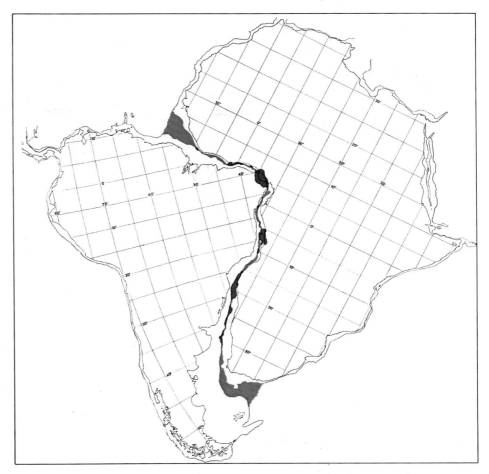

In 1964, British geophysicist Sir Edward Bullard suggested that a computer could be programed to find out how the Atlantic continents might have best fitted together 200 million years ago. The resulting map shows they meshed most precisely along their continental shelves some 3,000 feet below sea level.

of creeping ocean crust that bore the segments of ridge would scrape together and create earthquakes along their mutual boundaries. In Wilson's conception, however, one segment did not move away from the other. Both remained permanently in their place; the only movement was that of the sea floor itself as it spread relentlessly outward from the ridge. Only between the adjacent ridge offsets would the moving crust be shearing in opposite directions. Moreover, according to Wilson, the direction of shearing along such a fault would be opposite to that of a transcurrent fault *(page 90)*.

Wilson called these ridge-offsetting fractures transform faults, since they mark spots where the motion of the sea floor is transformed from a shearing movement between the offset ridge segments into a spreading movement away from the ridge. Proof of their existence, he wrote in 1965, "would go far towards establishing the reality of continental drift."

Such proof would be sought by Lynn Sykes, a young seismologist at the Lamont Geological Observatory. Already, Sykes had discovered that earthquakes on the East Pacific Rise were concentrated along the crest of the ridge, just as they were in the Atlantic; he had also found that earthquakes in the major fracture zones that offset the ridge were confined almost exclusively to the sections between the ridge segments—the very areas that Wilson had dubbed transform faults. Indeed, Wilson had noted, when introducing the concept of transform faults, that the newly identified features would explain Sykes's seismological findings.

When Wilson published his theory in July of 1965, Sykes was on an expedition in the Fiji Islands. He returned to Lamont during the winter, learned of the transform fault idea—and was unimpressed: Sykes had his doubts about sea-floor spreading. Then he studied the uncannily symmetri-

cal magnetic profile of the East Pacific Rise taken by the *Eltanin.* His conversion to Wilson's theory was immediate. "I saw the profile in June," he recalled later. "The next morning I went to work." He was determined to put the idea of transform faults to a rigorous test.

The key to proving the theory lay in determining whether the shearing movement of a transform fault was indeed opposite to that of a conventional transcurrent fault. A worldwide network of seismographic stations—installed to differentiate between earthquakes and underground nuclear explosions—had made it possible for scientists to compile precise data that could disclose the direction in which the earth's crust had moved during an earthquake. Most of this information was on hand at the Lamont laboratories, and Sykes pored over the microfilmed records in search of details about earthquakes that had occurred along the ridge fractures. Within a few days he had found 20 examples—and every one had been triggered by crustal slippage opposite to that predicted by traditional geology. Wilson had been right about transform faults. "Sea-floor spreading," said Sykes, "was the only thing that could produce those earthquakes."

Sykes first presented his findings to the Goddard Institute for Space Studies in New York in November 1966, and his material was reviewed later that month at a meeting of the Geological Society of America, held in San Francisco. At the same meeting, Frederick Vine delivered a paper rather hesitantly titled "Proof of Ocean-Floor Spreading?", but there were no questions in Vine's mind as he laid out the growing body of evidence that new sea-floor material was burbling constantly from the oceanic ridges and moving outward at a steady rate. Slowly, the scientific community was becoming convinced. By January of 1967 almost 70 abstracts on sea-floor spreading had been submitted for presentation at the upcoming April meeting of the American Geophysical Union, one of the most important annual scientific conclaves held in the United States. And the star attraction was Harry Hess, who was on the program to recount the background of his once-radical theory. Hess filled the largest auditorium available; those who could not find seats were happy to stand in the aisles to hear the man who had once described his ideas as geopoetry to avoid being laughed at.

What had happened, explained marine geologist Henry Menard of the Scripps Institution, was that scientists were saying to themselves: "My observations are not compatible with sea-floor spreading, and I shall prepare a critical demonstration that this is so and thus demolish this nutty idea so we can all get back to work." Instead, when they reexamined their data in the

A transcurrent fault, diagramed at left above, occurs between two tectonic plates that are sliding past each other; the friction causes earthquakes *(dots)* all along the fault. A transform fault *(above)* exists where sections of the Mid-Ocean Ridge are offset; as the sea floor spreads from the staggered segments, earthquakes occur where there is opposing movement along the fault.

light of the new theory, they were finding that sea-floor spreading was a thoroughly workable hypothesis.

And yet, concrete evidence remained tantalizingly elusive—until the advent of the *Glomar Challenger,* a highly sophisticated vessel designed by a California offshore-oil-drilling company to take core samples 30 feet long from previously unreachable depths of the ocean floors. In 1968, the ship began a series of voyages as part of a joint project involving the Lamont Geological Observatory and other leading oceanographic research centers. Boring through ocean sediment to the underlying crust with a drill string—a series of linked pipes—almost four miles in length, the *Glomar Challenger* took 2,509 feet of core samples from 17 widely spaced holes on each side of the Mid-Atlantic Ridge along lat. 30° S. and also along one of the magnetic lines that showed a good reversal pattern. It was a technological feat picturesquely compared to trying to drill holes in the sidewalks of New York from the top of the Empire State Building using a strand of spaghetti.

As they analyzed the fossil skeletons of marine organisms embedded in the cores, scientists found that the age of the oldest sediment in each core—that immediately on top of the basalt basement—was directly related to the distance of the hole from the ridge: the farther from the ridge, the older the fossil remains. This indicated a spreading rate of about three quarters of an inch per year on each side, which exactly agreed with the rate previously calculated by dating pole reversals and matching them to the magnetic stripes on the ocean bed. "Proof of sea-floor spreading," said expedition co-leader Arthur E. Maxwell of the Woods Hole Oceanographic Institution, "seemed to be in our hands."

The development and confirmation of the theory of sea-floor spreading was in itself a remarkable scientific bench mark. It would have been a fitting climax to a decade of extraordinary scientific advancement; in fact, there was much more to come, and it came with breathtaking speed. **Ω**

PROBING THE DEEP IN A PYGMY SUB

One of the best-known names in oceanographic research is Alvin; there is no surname, for Alvin is a homely little submersible built for the U.S. Navy in 1964, and operated ever since by the Woods Hole Oceanographic Institution. The *Alvin's* exceptional capabilities for deep-sea research have enabled scientists to observe and record evidence of sea-floor spreading at first hand.

Conventional submarines can dive no farther than about 1,000 feet because of the crushing water pressures at such depths. But the two-inch-thick sphere of titanium alloy that surrounds the *Alvin's* personnel compartment permits descents of more than 13,000 feet without danger to the pilot and two passengers. The battery-powered submersible's small size allows it to maneuver easily among the crags and chasms of the underwater terrain (at a modest top speed of 1.5 knots) while the mechanical arm plucks samples from the sea floor, and cameras and instruments record a wealth of observations.

Although self-sufficient during deep-sea dives, the *Alvin* is dependent on a mother ship, the *Lulu,* when on the surface. Nudging between the *Lulu's* twin hulls, the *Alvin* is hoisted to the main deck for servicing and the journey to the next dive site.

The *Alvin* has had moments of glory outside oceanography. In 1966, it located a U.S. Air Force hydrogen bomb that had tumbled to the ocean floor off the coast of Spain after a midair collision.

But the tiny vessel gained lasting scientific fame during the 1974 explorations of the Mid-Atlantic Ridge known as Project FAMOUS (French-American Mid-Ocean Undersea Study). Making repeated dives to a depth of 8,700 feet along a section of the ridge southwest of the Azores, the *Alvin* and two French submersibles gathered samples of volcanic rock and amassed some 100,000 photographs, many of them showing extruded lava formations. Delighted geologists were thus able to see, as well as analyze, the actual sites where volcanic activity is constantly adding to the sea floor along a midocean ridge.

A cutaway diagram reveals the compact design of the 25-foot-long minisubmersible *Alvin.* In an emergency, the sphere housing the three-man crew can jettison the rest of the hull and float to the surface *(inset).*

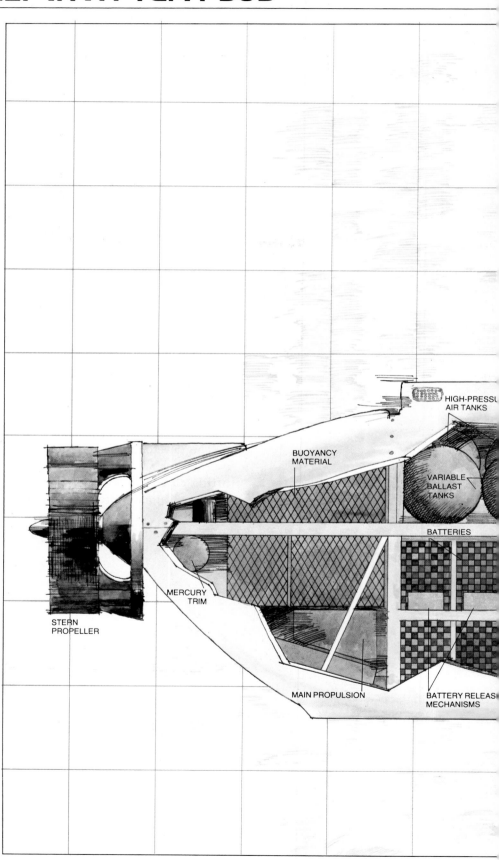

HIGH-PRESSURE AIR TANKS

BUOYANCY MATERIAL

VARIABLE BALLAST TANKS

BATTERIES

MERCURY TRIM

STERN PROPELLER

MAIN PROPULSION

BATTERY RELEASE MECHANISMS

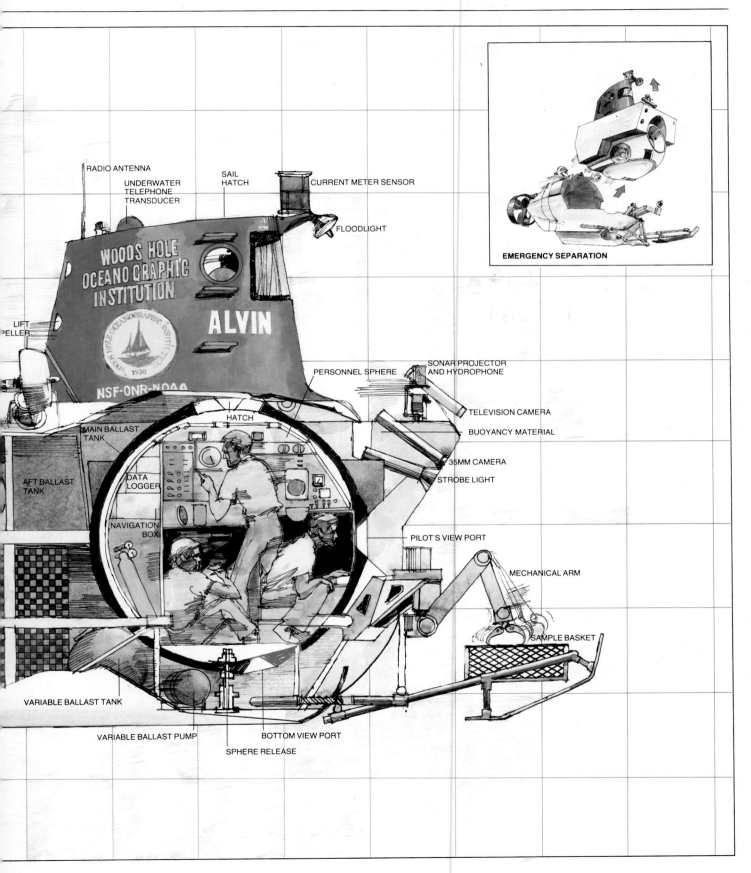

RADIO ANTENNA

UNDERWATER
TELEPHONE
TRANSDUCER

SAIL
HATCH

CURRENT METER SENSOR

FLOODLIGHT

WOODS HOLE
OCEANO GRAPHIC
INSTITUTION

ALVIN

NSF-ONR-NOAA

LIFT
PELLER

PERSONNEL SPHERE

SONAR PROJECTOR
AND HYDROPHONE

TELEVISION CAMERA

BUOYANCY MATERIAL

HATCH

MAIN BALLAST
TANK

35MM CAMERA

STROBE LIGHT

AFT BALLAST
TANK

DATA
LOGGER

NAVIGATION
BOX

PILOT'S VIEW PORT

MECHANICAL ARM

SAMPLE BASKET

VARIABLE BALLAST TANK

VARIABLE BALLAST PUMP

BOTTOM VIEW PORT

SPHERE RELEASE

EMERGENCY SEPARATION

93

Technicians inspect and service the *Alvin* on the main deck of the mother ship *Lulu (left)*. Guided by an exhaustive check list, they perform scores of tasks, including recharging the *Alvin's* batteries, checking the cameras and video-tape equipment, and replenishing the oxygen supply for the next dive.

Having lowered the *Alvin* from the *Lulu's* deck on a platform, crewmen ready lines to guide the minisub out from between the twin hulls of the mother ship. When the platform is lowered still farther, the *Alvin* will be afloat and free to move away for a dive.

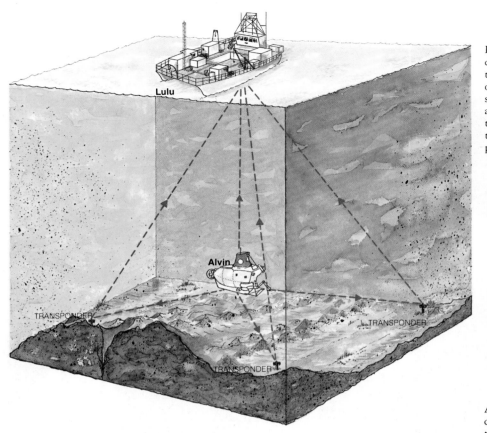

In the inky darkness of the sea floor two miles deep *(diagram),* the *Alvin* depends on the *Lulu* to keep track of its position. Sound impulses, or pings, emitted by the sub travel to the mother ship both directly and by way of transponders anchored to the sea floor. By comparing the times taken by each ping to reach the ship along the various paths, the *Lulu's* computer can plot the *Alvin's* location to within 20 feet.

As its ballast tanks fill with water, the *Alvin* sinks toward the ocean floor *(right, inset).* When the sub is on the bottom, the pilot *(right, center)* keeps watch through the main view port while two scientists behind him make observations through side view ports.

An oceanographer in the *Lulu's* control room confers with the *Alvin's* pilot by acoustic telephone. As a safety precaution, mother ship and submersible communicate at least once every half hour during a dive.

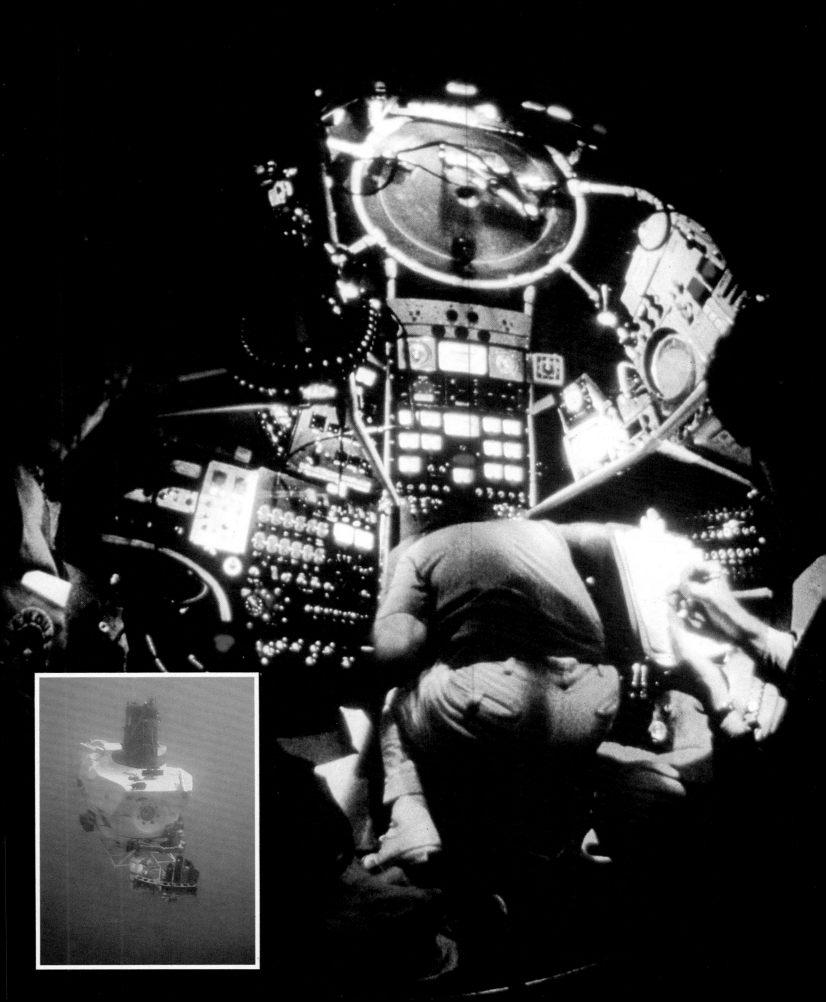

Lava in Haystack Formation

Lava Lining a Rift

Pillow Lava

A number of photographs taken by the *Alvin's* cameras in 1974 document the strange formations found on the ocean floor near the Mid-Atlantic Ridge. Formed as molten lava hardened in the frigid ocean depths, these extrusions provide convincing evidence of the nature of sea-floor spreading.

The *Alvin's* stately progress across the sea floor at a depth of 10,000 feet is captured by a remote-control camera. The mechanical arm extending from the bow is poised to collect a rock sample for deposit in one of the numbered bins of the hexagonal sample basket.

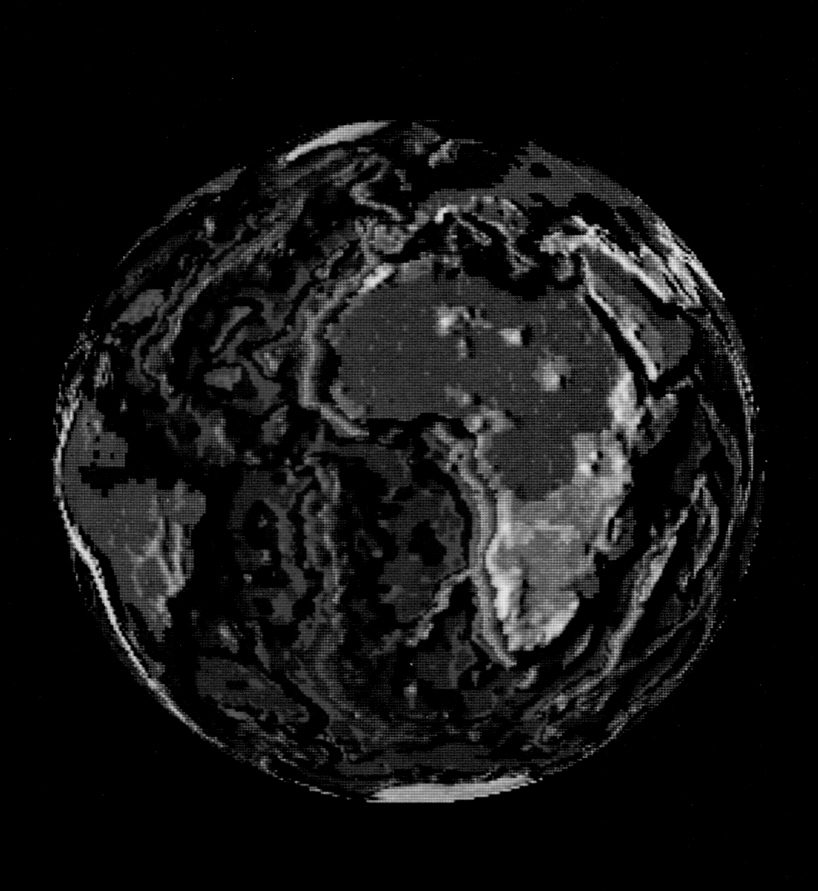

Chapter 4

THE RISE OF A NEW ORDER

It was as if we have been walking the deck of a ship, eyes down to study the deck, and never looked up to see that the ship itself was moving." Thus would Tuzo Wilson characterize the great awakening that swept the earth sciences as the revolutionary idea of spreading sea floors took hold.

Revolution does not come easily to the sciences, each of which depends on a body of specialized knowledge accumulated during many years of laborious experimentation, observation and documentation. Thomas J. Kuhn, a historian of science, has observed that, as time passes, more and more theories in a particular field come to be regarded as reliable and are no longer subjected to rigorous examination. Taken together, they form what Kuhn calls the science's paradigm—a consensus of scholars and researchers that becomes their doctrine. In the normal course of events, orthodox scientists go about the business of filling in gaps in their paradigm, framing their questions in terms of the paradigm's assumptions, seeking details that confirm its accuracy and efficacy. New theories that fit the paradigm tend to find quick and easy acceptance; those that contradict it are often deemed invalid and discarded out of hand. Thus, while science would undoubtedly be chaotic without such an orderly framework, the paradigm can, and often does, become a crutch—a comfortable substitute for critical thought.

Alfred Wegener ran a double risk when he advanced his theory of continental drift; not only did he challenge geology's paradigm of a cooling and contracting earth, but he did so from the vantage of an outsider: He had been trained as a meteorologist and astronomer. Defenders of orthodoxy naturally took deadly aim at the upstart. The eminent geophysicist Harold Jeffreys, for one, expressly proclaimed that Wegener's theory violated the laws of physics. Continents, said Jeffreys, simply could not plow around in the earth's mantle, which seismic data showed to be rock-solid.

Jeffreys' dismissal of continental drift was totally consistent with the consensus of scholars and sufficient to lead most of his colleagues to reject the theory (if they needed any nudging). Unfortunately, his facts were wrong: Later research would show that the mantle was not rigid after all.

Not until the 1960s did the paradigm begin to weaken. By then, geologists, oceanographers, geophysicists, seismologists and paleomagnetists were marshaling all sorts of facts that were at variance with the constructs and assumptions that had served for so many years. As the decade reached its midpoint, a few scientists were beginning to glimpse behind the damaged façade of the old patterns of thought an entirely new concept that could put sea-floor spreading and continental drift into a global framework.

This topographical map—generated by a computer using data from ocean- and land-based surveys—reveals the remarkable similarities between the sinuous track of the Mid-Atlantic Ridge and the edges of the African and American continents. The new vision of the earth's dynamics that emerged in the 1960s explained the processes responsible for such graphic parallels.

Tuzo Wilson was the first to sketch the shape of the future. Writing about transform faults in 1965, he suggested that the strange geological features could be explained as evidence for a network of "several large rigid plates" that were constantly in motion and that together made up the surface of the earth. In an accompanying map he showed how the worldwide pattern of midocean ridges, transform faults, and volcanically active mountains and island chains could be interpreted as the boundaries of these mobile plates. Wilson was more intent on explaining his transform fault concept than on enunciating a new global theory, and he did not elaborate on the moving-plates theme.

Dan P. McKenzie, a young English geophysicist, followed up on Wilson's rigid-plates idea in 1967, when he and a colleague, Robert L. Parker, published the suggestion that the earth's well-defined earthquake zones mark the boundaries of a number of rigid "paving stones." Analysis of earthquakes around the Pacific Ocean revealed a consistent direction of crustal movement; large segments of the ocean floor seemed to be both rigid and on the move. These paving stones, wrote McKenzie and Parker, interlock without gaps to form a global mosaic of plates.

The notion raised any number of knotty questions. If there were plates, and if they were moving, how and why did they move? How could their movements be predicted? What were the relationships between the shifting of the plates and the geological features of the earth?

The challenge of explaining how a huge, irregularly shaped plate might move across the surface of the earth did not go unmet for long. A Princeton geophysicist named W. Jason Morgan, who had been working independently on the idea of rigid plates ever since he became interested in sea-floor spreading through contact with fellow faculty member Harry Hess, discerned the answer in mathematics. During the 18th Century, Swiss mathematician Leonhard Euler had shown that when a section of a spherical surface is moved across the sphere, it necessarily rotates as it slides. Euler worked out a theorem for calculating the location of the axis of the rotation of the segment, an axis that always passes through the center of the sphere. Applying Euler's theorem to the supposed movements of plates on the earth, Morgan developed a way to use the direction of Tuzo Wilson's transform faults along the boundary of a plate to locate its axis of rotation. Once he knew that, he could calculate the direction and relative speed of the movement of any segment of the plate based on its distance from the axis.

To check his findings, he made a careful study of the magnetic stripes in the sea floor on either side of the Mid-Ocean Ridge, the great underwater mountain range that girdles the globe. If a plate were rotating, the edges of the plate close to the axis of rotation would move more slowly than the edges farther from the axis; thus, the stripes should get progressively wider relative to their distance from the axis. Morgan found that the rate of sea-floor spreading along the Mid-Atlantic Ridge did in fact vary according to the distance from the axis of rotation he had located; indeed, it did so with geometrical precision. Thus encouraged, he constructed a map indicating that the surface of the earth was divided into six large plates and about 12 smaller "subplates."

News of Morgan's work spread through the scientific community long before his paper was published, and dominated the conversations of the scientists who gathered for the April 1967 annual meeting of the American

Geophysical Union in Washington, D.C. The discovery of precise mathematical support for an idea that for years had been deemed mere fantasy lent an uncharacteristic sense of anticipation to the proceedings. Walter Alvarez—who was to become a professor of geology at the University of California at Berkeley—attended the sessions as a graduate student and later recalled "the atmosphere of electrical excitement running through the whole meeting." At one point, a professor gripped him by the arm and exclaimed, "We are witnessing a revolution!"

Morgan's paper, "Rises, Trenches, Great Faults, and Crustal Blocks," was published in the *Journal of Geophysical Research* in March 1968. As he explained his assumptions in his opening summary, "The earth's surface is considered to be made of a number of rigid crustal blocks. It is assumed that each block is bounded by rises (where new surface is formed), trenches or young fold mountains (where surface is being destroyed), and great faults."

Three months later, Xavier Le Pichon, a French oceanographer working at the Lamont Geological Observatory, drew on the work of Morgan, McKenzie and Parker and the geophysical data on file at Lamont to produce a masterly look at the history of major plate movements. Using Morgan's geometrical technique, Le Pichon showed how the six major plates had moved to open the Pacific, Arctic, Atlantic and Indian Oceans. He also confirmed Morgan's assumption that the surface of the earth is divided into a small number of rigid blocks in relative motion with respect to one another, adding a major point: "All movements are interrelated so that no spreading mid-ocean ridge can be understood independently of the others. Thus any major change in the pattern of spreading must be global." Even more arresting than his analysis of what had happened in the past was Le Pichon's willingness—with the limited amount of data available to him—to predict the nature of plate movements at distant locations. When his predictions were shown by later observation to be correct, the idea of moving plates had passed a critical test.

Before the year was out, three seismologists at Lamont—Bryan Isacks, Jack Oliver and Lynn Sykes—announced the results of another test. In a paper published in the *Journal of Geophysical Research* in September 1968, they enthusiastically pointed out that worldwide seismological evidence entirely supported the moving-plates hypothesis. Shallow earthquakes characterized the ocean ridges and transform faults; deep earthquakes occurred only where plates were being consumed at the ocean trenches. Furthermore, the direction of earth movement in earthquakes was consistent with Le Pichon's mathematical predictions of plate movement.

Included with the findings of Isacks, Oliver and Sykes was a spectacular map prepared by two other Lamont seismologists. They had painstakingly charted all the earthquakes that had been recorded between 1961 and 1967; the resulting map showed a spidery dotted line of earthquake epicenters that traced the path of the Mid-Ocean Ridge around the world, faithfully following the offset fractures along transform faults. In dramatic contrast, thick swarms of dots clustered over the ocean trenches and plate-collision zones. A wall-sized version of the map was hung in a hall at Lamont, and for weeks thereafter researchers spent their coffee breaks standing in the hallway gazing at the wondrous patterns. There, for everyone to see, were moving plates in action.

A skewed stretch of railroad track in south-
central Alaska testifies to the violence of
the 1964 Good Friday earthquake. Massive land
movements in a 100,000-square-mile area
were triggered when the advancing Pacific Plate
abruptly unlocked from the North American
Plate and plowed farther beneath it.

It was left to Isacks, Oliver and Sykes to coin a name for this new and increasingly useful concept, and they did so in the 1968 paper. They borrowed a term that had been used for more than 100 years to describe such dynamic geological processes as mountain building and folding—tectonics, from the Greek *"tekton,"* meaning "builder." To signify the novelty and scope of the concept, they proposed the term "new global tectonics," which popular usage soon amended to "plate tectonics."

Jack Oliver later recalled the excitement he and his colleagues had felt as they wrote the paper that did so much to confirm this revolutionary theory. "We sensed," he wrote, "we were involved in a major upheaval in the earth sciences in general, and in seismology in particular. Even if time should prove the entire concept in error, I am sure I would continue to think of those years as highlights of my scientific career, for surely the events transpiring then marked the start of a new era in the study of the earth."

Meanwhile, there were matters still pending from the former era. Even as some geophysicists were forging ahead into the strange new territory of plate tectonics, other scientists remained dubious about continental drift. But two significant discoveries cleared away much of the resistance.

Geologists from the Massachusetts Institute of Technology and the University of São Paulo in Brazil had decided in the mid-1960s to use radioactive isotope dating techniques to test the century-old observations of Alexander von Humboldt, the German scientist who had detected striking similarities between rock formations on opposite sides of the Atlantic. When they compared the rock strata on the west shoulder of Africa with those on the east coast of Brazil, they found that the ages and composition of the various layers matched precisely, indicating that the rocks were a part of the same formation. Of particular interest was a distinct boundary be-

Apartment buildings lie toppled like dominoes after an earthquake that ravaged the city of Niigata, Japan, in June 1964. Because of their proximity to a deep-sea tectonic battle line known as the Japan Trench, the Japanese home islands have endured many such cataclysms.

tween a two-billion-year-old geological province—a large area marked by similar geological history—in Ghana and a 600-million-year-old province beginning in a neighboring region in what is now Benin. The boundary headed into the ocean near the city of Accra, and geologists calculated that if Brazil had indeed been joined to Africa, the boundary would reappear on the northeast coast of Brazil, near the town of São Luís. "To our surprise and delight," said Patrick M. Hurley, leader of the project and a former skeptic in the matter of continental drift, the "boundary line lay exactly where it had been predicted."

Hurley wrote of his findings in 1968, and the following year paleontologists on an expedition to Antarctica made what was described as "one of the truly great fossil finds of all time." On a north-facing cliff called Coalsack Bluff, they came across the bones of a sheep-sized reptile called *Lystrosaurus,* the first vertebrate land animal ever found in Antarctica. *Lystrosaurus* was known to have lived in Africa, India and China between 180 and 225 million years ago. "The fully developed presence of *Lystrosaurus* in Antarctica," reported Edwin H. Colbert of the American Museum of Natural History in New York, "indicates that Antarctica and southern Africa were joined along a broad front. The same is true for peninsular India, making of them essentially a single land."

These geological and fossil discoveries were clear confirmation that the modern continents were once joined and had later drifted apart. But by the time the two proofs came to light, the plate tectonics revolution had already gathered remarkable momentum, capturing the imaginations of earth scientists throughout the United States and Europe. The extent of the new theory's influence did not become fully apparent until a convention of geophysicists was held in Moscow in 1971. Only a few years earlier, the overwhelming majority of the scientists attending a similar geophysical conference had considered the idea of continental drift to be sheer heresy. Now almost every Westerner at the convention not only accepted the movement of continents but had moved on to become an advocate of plate tectonics. It was an extraordinary turnabout.

Not long afterward, the theory of plate tectonics made an unusual convert in dramatic circumstances. The American astronaut Harrison Schmitt, orbiting the earth in the *Apollo 17* spacecraft in December 1972, looked down on the northeast corner of the African continent and told Mission Control: "I didn't grow up with the idea of drifting continents and sea-floor spreading. But I tell you, when you look at the way the pieces seem to fit together, you could almost make a believer out of anybody."

As refined by the efforts of earth scientists around the world, the theory of plate tectonics holds that the earth's outer shell, called the lithosphere, is divided into nine large plates and a number of smaller plates. Except for the oceanic plates—the Pacific, Nazca and Cocos—the major plates are named after the continents embedded in them: North America, South America, Eurasia, Africa, India-Australia and Antarctica. Most plates comprise both continental and oceanic crust, although one of the largest (the Pacific Plate) is almost entirely oceanic, and one of the smallest (the Turkish Plate) is entirely land.

The continents themselves bear little relationship to the size or shape of the plates; they simply ride on them like logs stuck in an ice floe. In contrast to the ocean floors, which are constantly being consumed at the sub-duction zones and regenerated at the Mid-Ocean Ridge, the continents are permanent features of the earth's crust: Because they are lighter and thicker than the basalt of the ocean crust, they never sink. When, inevitably, continents carried on converging plates collide, the edges crumple and the plates become jammed, their movements temporarily halted.

About 60 miles thick, rigid and immensely strong, the plates are floating on a soft, hot, semiviscous layer of the earth known as the astheno-sphere, whose presence has been deduced from sophisticated analyses of seismic waves. Beneath the asthenosphere lies the earth's deep mantle, the mesosphere. High temperatures and tremendous pressures on the topmost part of the asthenosphere cause it to deform and even flow plastically, thus permitting the plates—driven by forces that are not yet fully under-stood—to inch along on their endless journeys around the surface of the earth, moving no faster than human fingernails grow.

The subtleties of this crustal pushing and shoving are countless: None of the plates is a simple rectangle, neatly emerging along one edge, sliding past other plates along opposite sides and diving back into the mantle along the fourth edge. Instead, the boundaries are made up of a complicated series of ridges, where adjoining plates are pulling apart from each other; trans-form faults, where they are sliding past each other; and subduction trench-es, where two plates collide and one is thrust under the other and descends into the mantle.

Not all the plates are consumed in trenches along their leading edges, however. The African and Antarctic Plates are almost totally surrounded by sea-floor-forming ridges, and are nowhere being destroyed; yet neither continent shows any signs of compression. The only explanation for the apparent contradiction is that these plates are increasing in size; the ridges themselves must be moving outward from the continents, decreasing the size of plates elsewhere.

Fundamental to the theory of plate tectonics is the assumption that while all the plates seem to be moving at different relative speeds—ranging from a fraction of an inch to a maximum of five inches a year—the whole jigsaw puzzle of plates is interlinked. No one plate can move without affecting others, and the activity of one can influence another thousands of miles away: The Atlantic Ocean could not be getting wider—as it is with the spreading of the African Plate away from the South American Plate—if the Pacific sea floor were not being consumed in deep oceanic trenches faster than it is created at the Pacific ridges. The plates move rapidly by geologi-cal standards: Two inches per year—to pick a typical speed—adds up to 30 miles in one million years. It took only 150 million years for a mere fracture in an ancient continent to turn into the Atlantic Ocean.

The driving force behind the process of plate tectonics is heat. Generated deep inside the earth by radioactive decay, the heat reaches the surface primarily along the Mid-Ocean Ridge. Lateral tension created as the plates are pulled apart along the ridge reduces the vertical pressure on the underly-ing hot magma—liquid rock—and allows it to well up from the mantle. The magma oozes forth as lava to fill the crack continuously generated by plate separation, cools to a rich basalt and welds itself to the trailing edge of the plate on each side of the ridge, forming a new ocean-floor crust. Where the plates are moving rapidly apart, as along the East Pacific Rise,

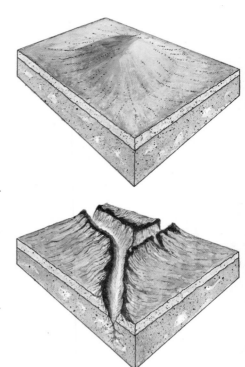

The unending rearrangement of the world's land masses involves the occasional sundering of continents. Such a division begins when magma heats continental crust and forces up a dome *(top)*, which then cracks open to form a Y-shaped triple rift *(bottom)*. At right, a view from space shows a classic triple rift between North Africa and Arabia: Two of the arms are the Red Sea and the Gulf of Aden; the third is the Great Rift Valley, which is slowly separating two sections of the African continent.

no rift valley is formed in the ridge: The new sea floor simply migrates along sloping flanks on either side of the rise. Where spreading is slower, as along the Atlantic and Indian Ocean ridges, precipitous scarps are formed on each side of a deep rift valley marking the center of the ridge.

Transform faults, where newly formed plates slide past each other, are normally aligned with the direction of plate movement. If the faults are offset at an angle, either a deep chasm is formed as the plates pull obliquely apart—the basin of the Dead Sea might have been formed in this way—or the plates converge obliquely and compress, folding mountain ranges or shearing along new faults.

The most famous and most visible transform fault in the world is the San Andreas Fault, stretching for 1,000 miles from Cape Mendocino in northern California through western California to the East Pacific Rise beneath the waters of the Gulf of California. The Pacific Plate, on the west side of the San Andreas Fault, moves northwest at the rate of about two inches every year relative to the American Plate on the east side of the fault. Much crushing and grinding takes place as the two plates move past each other; when sections of the plates become locked, stress builds up until the friction is relieved by a major earthquake. This is what happened at 5:12 a.m. on April 18, 1906. In less than a minute, one plate lurched 20 feet northward along a 270-mile stretch of the fault, releasing pent-up energies that had been growing for more than a century, and causing the catastrophic San Francisco earthquake.

While new ocean crust is constantly being generated, old crust must either be destroyed or reduced at the same rate. Therefore, the plates, emerging along midocean ridges, sliding over the asthenosphere and grinding past other plates along transform faults, are almost all headed on a collision course. Nowhere is the collision more spectacular than when two continents, carried on converging plates, ram into each other, crumpling and folding under the enormous pressure, and creating great mountain ranges. A majestic example of a continent-to-continent collision is the snow-capped Himalayas, the highest mountain range in the world. This string of towering peaks is still being thrust up, as India, embedded in the Indo-Australian Plate, continues to crunch relentlessly into Tibet, on the southern edge of the Eurasian Plate. The European Alps have been formed in similar fashion, starting some 80 million years ago when the outlying continental fragments of the African Plate collided with the Eurasian Plate. Unrelenting pressure between the two plates continues, and is slowly closing up the Mediterranean Sea.

When two oceanic plates collide, the result is less obvious—but no less dramatic. The younger of the two plates will be lighter (plates grow denser as they cool). In consequence, it rides over the edge of the other plate, which then bends and plunges steeply through the asthenosphere. As the older, heavier plate descends into the earth, it forms a trench that can be as much as 60 miles wide, more than 1,000 miles long and several miles deep. The Marianas Trench, where the great Pacific Plate is descending, is the deepest sea floor in the world; curving northward from near the island of Guam, its bottom lies some 36,000 feet below the surface.

If the descending plate is carrying a continent, the arrival of the land mass over the trench causes mammoth complications. The lighter continental material cannot sink, but it dives into the trench behind the oceanic

Many of the essential clues to the tectonic processes that shape the earth were found beneath the sea, in samples painstakingly drawn from the deep-ocean floor. The effort to find the evidence was spearheaded by an innovative drilling ship called the *Glomar Challenger*.

Commissioned in 1967 by a consortium of oceanographic institutions, the *Glomar Challenger* presented a unique design problem. While seagoing oil-drilling platforms anchor in the relatively shallow waters above the continental shelves, the *Challenger* would be working in 20,000-foot-deep waters; obviously, other means would be required to hold the ship in place while its drill string—sections of five-inch pipe coupled to form a flimsy shaft almost four miles long—bored holes thousands of feet deep into the sea floor.

For starters, the designers made sure the 400-foot-long *Challenger* was heavy for its size and innately stable. Because some heaving in the seas was inevitable, they provided telescoping sections of drill pipe to hold the bit steady as it gnawed into the sea floor. But the vessel's most innovative accouterment was a computer-controlled positioning system. By compensating for wind and currents with main and thruster propellers, the system held the ship nearly motionless over the drill hole, even in storms so severe that the main deck was awash.

Core samples—cylinders of sea-floor rock and sediment up to 30 feet long—were the scientific reward for this ingenious engineering. Within 13 years of its first expedition in 1968, the *Challenger* had retrieved almost 50 miles of cores from more than 500 sites around the world. The samples provided positive proof of sea-floor spreading, yielded data on the marine life of past eras and helped scientists comprehend the mechanics of sea-floor formation. So many scientists, from volcanologists to biologists, were interested in doing their own analyses that it was impossible to circulate the cores fast enough—testimony to the unparalleled utility of the *Glomar Challenger* as a tool of deep-sea research.

The drilling ship *Glomar Challenger* stands off the coast of Brazil during a 1980 expedition to gather sea-floor samples from the South Atlantic. Its 142-foot derrick can support a million pounds of drill pipe.

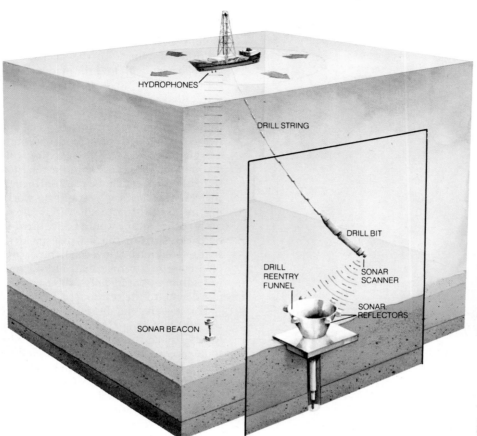

The *Glomar Challenger's* dynamic positioning system monitors a sonar beacon on the sea floor, detects the ship's drift and activates propellers to hold the ship's position above its drill site. A 16-foot-wide funnel *(inset)* equipped with sonar reflectors enables the *Challenger* to return its drill to the miles-deep hole.

Workers atop the *Glomar Challenger's* drill platform use hydraulically powered tongs to link 90-foot sections of drill pipe. The tongs can screw two pipes together in 30 seconds; the sections are then lowered through the platform so the next length can be attached.

Spraying water over the platform, the drill pipe is hoisted aboard after its sample has been removed. Water trapped by a clogged drill bit rushed out as each length of pipe was unscrewed.

After retrieving a core barrel from inside the drill pipe, two workers *(lower right)* hold it steady while members of the *Glomar Challenger's* scientific team extract the 30-foot-long plastic tube that contains a sample.

In the *Glomar Challenger's* laboratory, a technician slices a core sample in half lengthwise with a diamond-tipped, water-lubricated saw. One half will be analyzed aboard ship; the other half will be preserved for future study.

After each sample has been meticulously catalogued *(above)*, small sections are washed, screened and dried to extract the fossil remains of single-celled animals. By identifying the fossils under a microscope *(right)*, paleontologists can estimate the age of the sea-floor layers from which they came.

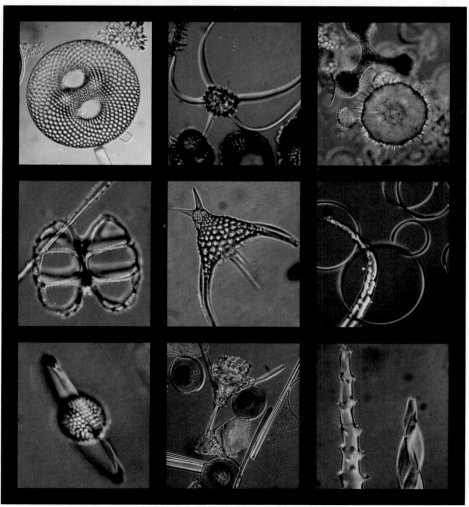

crust until it gets stuck. This crumples its leading edge into folded mountains and causes some of the oceanic crust to be deposited on top of the continent. Meanwhile, pressure builds up until the trench "flips," and the previously overriding oceanic plate dives under the continental plate. (This process may explain why most ocean trenches are found along the edges of the continents.)

As an undersliding plate tips down, its accumulated ocean-floor sediment—mostly eroded soil and skeletons of marine organisms—is scraped off against the front of the overriding plate, increasing its width and thickness. This is one possible explanation for the fact that the Andes, the long mountain range bordering the west coast of South America, appears to be growing higher. Perhaps sediment from the Nazca Plate, which is diving under South America in the Peru-Chile Trench, is scraping off on the granite roots of the Andes, adding thickness and buoyancy to the mountains so that they continually float upward more rapidly than they can be eroded by wind and rain.

If a trench has flipped because of the arrival of a continent, and the newly subducting plate also carries a continent, a collision of land masses is inevitable. When it happens, subduction terminates along the collision zone and the trench disappears, causing the neighboring plates to readjust their motions. An entire plate can disappear if its leading edge is being consumed faster than new crust is being added at the ridge on its trailing edge. If this occurs, the ridge is slowly pulled toward the trench and the whole plate is eventually drawn down into the mantle, necessitating a rearrangement of other plates and their boundaries.

The sinking of a plate into a trench is a long and complex journey, and its effects are apparent on the face of the earth. As the cold and stiff plate begins its descent, it generates an almost continuous series of earthquakes. On its way, the plate starts to heat up through a combination of friction, contact with the warmer material of the mantle, pressure, radioactive decay and energy released as the lithospheric materials are compressed into denser, more compact crystal structures. At a depth of about 40 to 80 miles, "differentiation" occurs, as lighter magmas are selectively melted and rise toward the surface, eventually intruding into the leading edge of the overriding plate, where they add material to the crust and build volcanoes above it. If the upper plate is oceanic, the volcanoes pile up until they poke through the surface of the ocean in a graceful arc—a distinctive configuration related to the curvature of the earth.

Island arcs are clustered like a necklace around the northern and western margins of the Pacific: the Aleutians, the Kuriles, Japan, the Ryukyus and the Philippines. To the south lie Indonesia, the Solomons, the New Hebrides and the Tongas. Their origin had puzzled scientists for centuries. Cartographers had charted the deep ocean trenches, seismologists had plotted earthquakes beneath the trenches and volcanologists had studied the overlying volcanoes, but these researchers had generally worked independently, unaware that the phenomena they studied were part of the same process. Now the ideas of sea-floor spreading and subduction finally explained why so many of the world's volcanoes are situated on the Pacific island arcs, the so-called Ring of Fire, where tectonic plates are being subducted beneath the ocean trenches.

The final phase of plate subduction begins after about 10 million years,

and at depths of as much as 450 miles, when the plate becomes so hot that it softens and ceases to generate earthquakes. The descent and the melting continue until, at some unknown depth, the plate blends with the surrounding mantle material.

The development of the theory of plate tectonics was a time of unparalleled excitement for those lucky enough to be participants; Tanya Atwater, a graduate student at the Scripps Institution in the late 1960s, later recalled an evening spent drinking beer with friends at a dance hall in La Jolla, California; the group watched with growing delight as Dan McKenzie sketched on a napkin an explanation of how plate tectonics accounted for the geometry of the San Andreas Fault. "It is a wondrous thing to have the random facts in one's head suddenly fall into the slots of an orderly framework," she recalled. "It is like an explosion inside. That is what happened to me that night. The simplicity and power of the geometry of those plates captured my mind and has never let go since. Sea-floor spreading was a wonderful concept because it could explain so much of what we knew, but plate tectonics really set us free and flying."

It is impossible to pinpoint the date of this revolution; different people endorsed the new theory at different times, depending on the information that was available to them, the nature of their specialties, even their personalities. But when the earth scientists had, in Atwater's phrase, been set free, the excitement was replaced with another kind of exhilaration—that of testing the new paradigm against new observations, experimentation and calculations, and finding that it held.

In the early 1970s, for instance, scientists mounted an ambitious research project to find out precisely what happens at the crests of the mid-ocean ridges. Maurice Ewing had discovered newly formed volcanic rock along the Mid-Atlantic Ridge as early as 1947, and his findings had been confirmed in 1961 by samples brought up by an oceanographic research ship from Woods Hole. In 1972 a French ship towed a camera-equipped sled along the ridge southwest of the Azores and photographed curious tubular extrusions of lava, known as pillow lava, or "toothpaste flows," that are formed when molten rock rapidly solidifies upon contact with sea water thousands of feet beneath the surface of an ocean. Samples recovered and hoisted to the deck of the ship popped and exploded in the reduced surface pressure, indicating the presence of trapped volcanic gases.

During the following year, in preparation for a Jules Verne-type expedition known as Project FAMOUS (French-American Mid-Ocean Undersea Study), Xavier Le Pichon made the first manned dive into the rift of the Mid-Atlantic Ridge in a research submersible, the *Archimède*. More than 7,000 feet below the surface, the *Archimède's* powerful searchlight illuminated a bizarre submarine landscape of bulbous pillow-lava formations festooning the sides of the rift in strange shapes, like wax overflowing from a candle. For Le Pichon, it was a spine-tingling moment: He was witnessing, with his own eyes, what brand-new ocean crust looks like soon after it bubbles up from within the mantle. The pilot of the craft set down on the slope of the rift, where Le Pichon "sampled a fairly large pillow while fishes kept poking at our port holes."

In the summer of 1974, two French submersibles and the United States Navy's undersea research vessel *Alvin* made a total of 44 dives to depths

Almost all the living things that populate the world's oceans are found within 700 feet of the surface; below that level, life-sustaining sunlight gives way to cold and darkness, and few creatures can be found. However, at more than 10 times that depth, near a series of undersea hot springs on the submarine ridges of the Pacific Ocean floor, unique lifeforms abound, singular products of the earth's tectonic movements.

Scientists in the 1960s predicted the existence of hot springs—geysers of magma-heated sea water—on the crests of midocean ridges. But it was not until 1977 that an expedition to a rift northeast of the Galápagos Islands located such a spring with an underwater temperature sensor. When scientists went down 8,500 feet for a closer look in the research submarine *Alvin*, their lights revealed a surprising profusion of giant, crimson-tipped worms, scampering white crabs and football-sized clams.

In later studies with the *Alvin* on the Galápagos rift and at scalding submarine geysers on the East Pacific Rise, scientists unraveled some of the mysteries of this deep-sea luxuriance. In place of the sunlight that fuels life in brighter regions, these undersea oases rely on the chemical energy contained in hydrogen sulfide, a foul-smelling gas carried out of the earth's crust by the mineral-rich springs and geysers. Bacteria—so numerous that they form thick mats on adjacent rocks—thrive on the sulfides and in turn serve as food for the larger animals that crowd around the springs.

Investigators believe such oases occur all along the ridges of the Pacific and Indian Oceans. Contemplating the extent of this potential habitat for uncatalogued species, one scientist said, "It's like discovering a new continent, with all its flora and fauna."

Black with suspended minerals, water superheated to 700° F. belches from the East Pacific Rise, south of Baja California. Bottles for water collection rest against a rocky chimney that was formed as upwelling minerals precipitated and hardened.

As the sample basket of the research submarine *Alvin* approaches, foot-long clams lie strewn across the ocean floor near the Galápagos rift. The giant clams are nourished by a rich seawater broth of bacteria and sulfides.

Water warmed by underlying magma seeps from a fissure on the Galápagos rift as the probe of the *Alvin* measures its temperature—cooled to about 65° F. by the sea water infiltrating the ocean floor. The water's milky color is caused by bacteria and sulfur compounds.

Flowering in total darkness on the Galápagos rift, the two-inch "dandelion" at left is actually a species of animal related to the Portuguese man-of-war. Its petal-like organs surround a central gas pouch that gives it buoyancy.

bacteria living inside the worms absorb
the sulfides and produce usable food.

of between 7,200 and 9,600 feet in the rift of the Mid-Atlantic Ridge. With clawlike mechanical arms, they collected one of the most remarkable sets of rock samples ever obtained from the sea floor and collated a mass of data—including detailed heat-flow and magnetic measurements and high-resolution photographs—that helped scientists further understand the enigmatic process of sea-floor spreading.

During their dives, cruising a few feet above the floor of the rift, American scientists on board the *Alvin* discovered more than 400 cracks running parallel to the ridge and ranging in width from a few inches to 40 or 50 feet. Some were half a mile or more long and perhaps as much as 50 feet deep. (While exploring one of the deeper fissures, the *Alvin* became wedged for a few heart-stopping minutes, and it took the pilot more than an hour to jockey his tiny craft out of the crevasse.) These cracks were important because they indicate that the plates are being pulled apart by some force, rather than being pushed apart. Indeed, in some of the narrower fissures, matching halves of the same lava pillows could be seen on opposite walls.

Even more than personal observation of the results of plate tectonics, scientists value the stern tests of mathematics—the checking of a mathematical model of a process against its actual workings. And such a test was applied to the very depths of the oceans.

As the growing oceanic plate moves away from the crest of a ridge, it cools and contracts, slowly sinking deeper into the mantle as it becomes denser. The process is so undeviating that there is a striking relationship between the age of the sea floor and the depth of water covering it. Dan McKenzie had theorized about this relationship in 1967, calculating that the overall profile of the ocean floor was simply a reflection of cooling and contraction. His work was followed by the studies of John G. Sclater—who, as a research student at Cambridge in 1965, had worked with both Harry Hess and Tuzo Wilson. In 1971, Sclater and his students at Scripps collected and plotted all the available data on the age and depth of the Pacific floor. Their results confirmed McKenzie's mathematical prediction, moving the Japanese geophysicist Seiya Uyeda to note: "It is a well-known fact of oceanography that the eastern Pacific region is shallow and the western region deep, and yet it is a thrill to observe that the plate model can explain it as a universal and inevitable consequence of a simple phenomenon like thermal contraction."

Not all of the checks of observation against theory worked so neatly. For example, most scientists who accepted the concept of plate tectonics believed that the basic sizes and shapes of the continents had been fixed relatively early in tectonic history. But some strange rocks first noticed in 1963 by Warren Hamilton seemed to indicate otherwise. Hamilton, a young researcher working out of the U.S. Geological Survey's Denver office, was studying the geology of a 1,000-square-mile section of western Idaho when he came upon rocks that had obviously originated in the ocean, on a volcanic island. Exploring further, he found that the entire western half of his area was peppered with oceanic rock—and this in a region some 400 miles inland from the Pacific.

The idea of mobile land masses had not yet gained a firm scientific foothold, so when Hamilton described his findings in a paper published later in the year, he prudently declined to commit himself to the obvious conclusion: A parcel of Idaho was made up of an island chain that had somehow welded itself to the North American continent. "I didn't quite have the guts to say it in print," he would recall. "I thought it; then I tiptoed away." Six years later, after finding similar evidence in California—and after watching the scientific community embrace the concept of plate tectonics—Hamilton at last spoke out. Parts of North America, he wrote, must have been formed as the continent moved through the ocean and bashed into wandering island chains.

Over the next few years, an accumulation of additional evidence indicated that as much as 25 per cent of North America had originated elsewhere, as ocean crust, volcanic islands or parts of other continents that had broken apart as a result of jostling among tectonic plates. Paleontologists in Canada and Alaska, for example, found the fossil remains of extinct one-celled organisms that had once inhabited the shallow waters off China, Indonesia and Japan. And paleomagnetic readings indicated that parts of Alaska had formed as far south as the Equator. Indeed, researchers have concluded that nearly half of Alaska is made up of some 50 distinct fragments that have become plastered onto the continent. And it is likely that many other parts of the world have been built up in a similar fashion.

The realization that continents have been formed in part by the coming together of random bits and pieces of rock masses—called microplates, though they do not move on plates of their own—explains the presence of such geological oddities as the oceanic rocks of Idaho, and the study of the histories and travels of these microplates soon became an interesting specialty of plate tectonics. Many researchers believe that near-coastal mountains such as the Andes and the Canadian Rockies were thrust up when microplates were pressed against larger continents. It is not known for certain just where all these mountain-building bits of terrain came from. Two scientists, Amos Nur and Zvi Ben-Avraham, have proposed that a number of them were remnants of a fractured former continent that the pair has dubbed Pacifica, but much more research will be required before the existence of such a hypothetical land mass can be confirmed.

Such puzzles, especially when compared with the far-reaching successes of the theory of plate tectonics, may reasonably be regarded as loose ends that are fascinating in detail but whose resolution is merely a matter of time. Meanwhile, scientists are left to ponder a more basic question: What keeps the plates in motion? Ω

EXPLAINING A DYNAMIC EARTH

The emergence of the theory of plate tectonics provided scientists with a unifying framework for understanding how and why the various features of the earth constantly change. In this new view of the earth, the seemingly immutable mountains, oceans and even continents are, in terms of geologic time, fleeting manifestations of a far grander reality.

The theory describes a perpetually changing mosaic of rigid segments, or plates, of the earth's outer layer, the lithosphere, which consists of the crust and the upper mantle. These plates, 30 to 90 miles thick, drift about at varying speeds and in different directions atop a hotter, softer layer, the asthenosphere.

As they move, they carry along great blocks of the earth's surface, which ponderously converge, separate and slide past each other.

Plate borders cut through continents and oceans and are concealed by them. But titanic geological events along these boundaries offer clues to their locations. Where plates converge, mountains and volcanoes are often found; where they pull apart, oceans are born; and wherever they grind against each other, they are racked by frequent earthquakes.

The sea-floor plates are continually being created at rifts along a global network of midocean mountain ridges. Here, basaltic magma from the earth's

molten interior regularly oozes onto the sea floor. Cooled by the water, it solidifies and is carried outward from both sides of the rift (*below*).

At the same time, the plates are being consumed in subduction zones, deep trenches where the slabs descend into the mantle. At depths of 40 to 80 miles, the crust begins to melt; the resulting magma rises through the overriding plate, forming a chain of volcanoes.

Although continental crust is five to 10 times thicker than oceanic crust, it is much less dense and thus resists subduction; when continents plow into each other, they simply fold and double into mighty ranges of mountains.

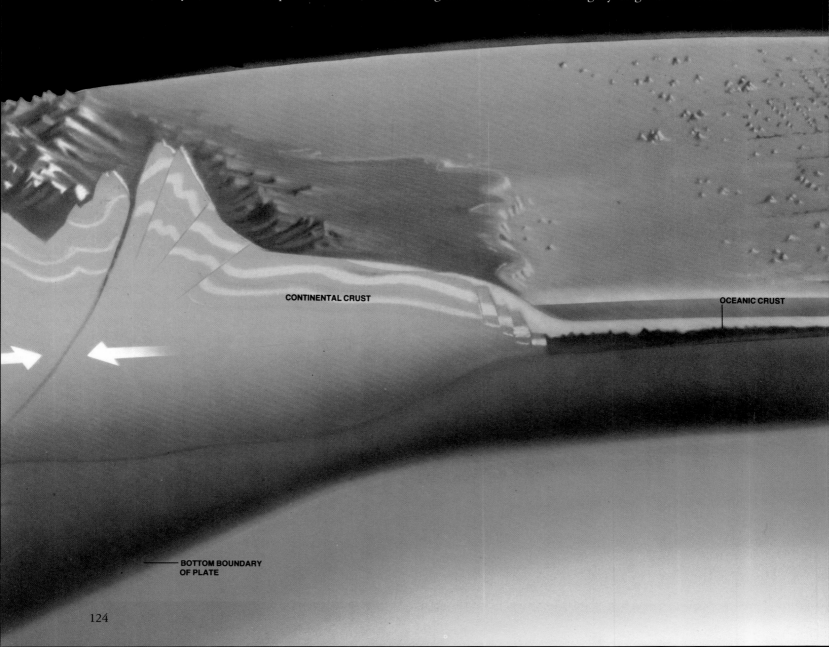

CONTINENTAL CRUST

OCEANIC CRUST

BOTTOM BOUNDARY
OF PLATE

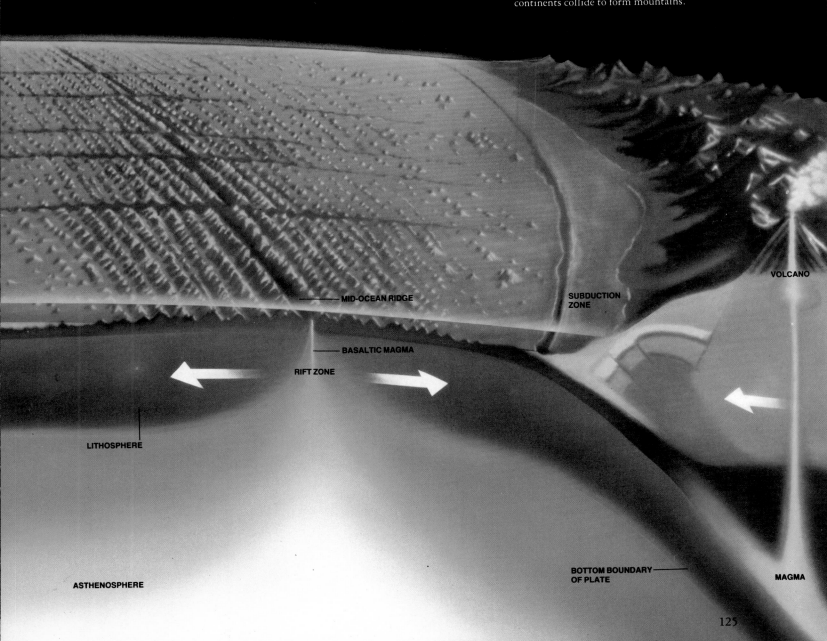

A cross section shows how tectonic processes continually renew and reshape the earth's surface. At right, sea floor created at an oceanic rift spreads away to a subduction trench, where the plate's descent into the hot asthenosphere is marked by volcanoes; at left, plates carrying continents collide to form mountains.

VOLCANO

MID-OCEAN RIDGE

SUBDUCTION ZONE

BASALTIC MAGMA

RIFT ZONE

LITHOSPHERE

BOTTOM BOUNDARY OF PLATE

MAGMA

ASTHENOSPHERE

When a tectonic plate carrying a continent splits apart, a new ocean basin is created. The process is believed to begin when the continental crust is arched, weakened and stretched by the heat from basaltic magma welling up from deep within the earth (right. top). The brittle crust fractures on each side of the stressed area, allowing sections to drop. The result is a rugged rift valley, slashed and scarred by faults and fissures in which eroded sediments accumulate (right. middle).

This early stage of ocean building can be seen in several parts of today's world, including the Baikal region of southeastern Siberia and the United States from western Utah to eastern California, an area known to geologists as the Basin and Range. But the most dramatic example is in the huge rift valley of East Africa, stretching between Ethiopia and Tanzania (pages 30-39). The volcanic peaks of Mount Kilimanjaro and Mount Kenya are by-products of this rifting process.

As the rift widens, the floor eventually drops low enough for the sea to invade the gap between the two land masses (right. bottom). This stage is typified by the Red Sea, a widening ocean basin located where the Arabian Peninsula was severed from Africa by the pulling apart of the African and Arabian Plates.

As the young ocean widens and matures, the rift develops a ridge of lava mountains on either side (below). The Mid-Atlantic Ridge, for example, rises where the American continents are separating from Europe and Africa. The Mid-Ocean Ridge is one of the earth's most dramatic topographical features—a continuous range of undersea mountains more than 12,000 feet high and 1,200 miles wide winding through 40,000 miles of the world's oceans.

A jagged oceanic ridge gradually evolves along the rift zone, shown here at the center of a vast, mature ocean tens of millions of years old. Since new lava is added to the separating plates in equal increments, sea-floor spreading is normally symmetrical.

OCEANIC CRUST

Magma welling upward from the interior of the earth bulges and weakens the crust of a continent that is about to be split in two.

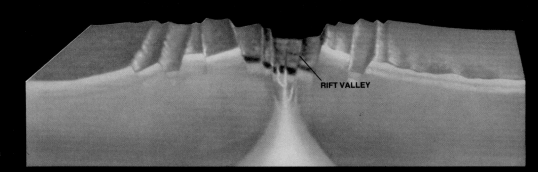

MAGMA

RIFT VALLEY

Sections of the heated and elevated continental crust fracture, slide away from each other and drop to create a rift valley.

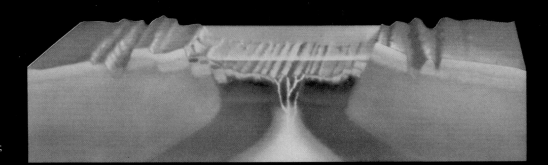

The continental sections move in opposite directions as lava spreads outward from the rift; sea water eventually fills the widening gap.

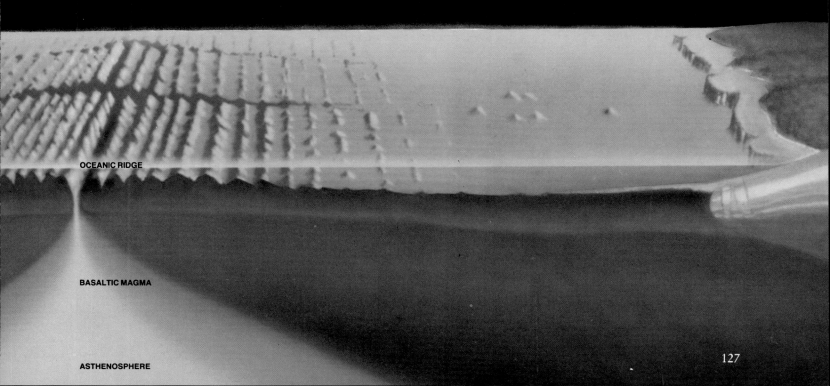

OCEANIC RIDGE

BASALTIC MAGMA

ASTHENOSPHERE

127

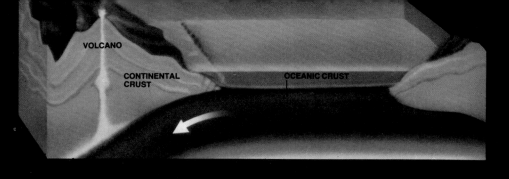

When a subduction trench forms under the edge of a continent, the melting of oceanic crust produces volcanoes on the overriding plate.

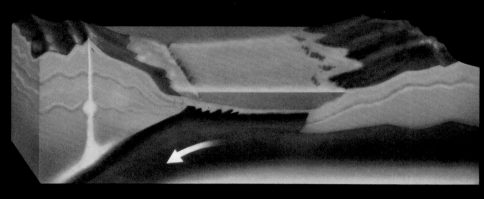

As the ocean gradually narrows and another continent approaches, the subduction process accelerates, increasing the accompanying volcanic and earthquake activity.

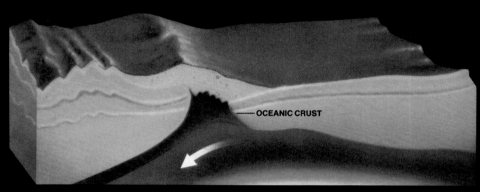

Finally, the ocean disappears and the continents collide. Deformed shreds of oceanic crust and wedges of sediment are plastered against the edges of the continents.

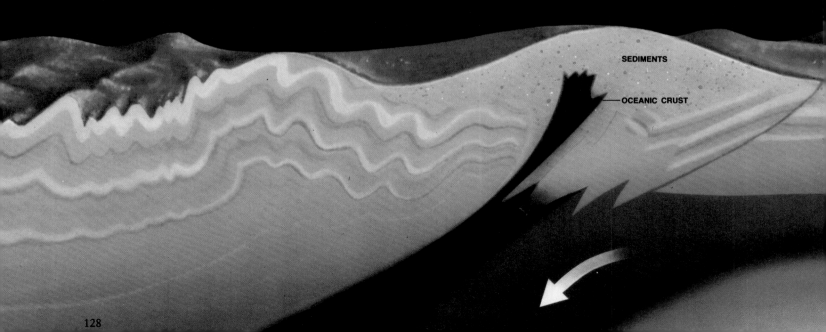

Building Continents
and Folding Mountains

Although the origin of the world's land masses remains a matter of scientific conjecture, the birth of continents may begin when two plates carrying ocean basins converge. The awesome tectonic forces drive one of the plates downward through a subduction trench; as it begins to melt, magma percolates upward and erupts along the edge of the overriding plate to form a chain of volcanic islands (*right*). Upwellings of magma that do not reach the surface harden into lighter, granitic, continental material. Continuing volcanism and erosion extend the islands laterally, perhaps until they attain continental dimensions.

However formed, a continent moves ceaselessly on its underlying plates and occasionally converges on another (*left*). The oceanic crust between them is slowly subducted until the continents collide and fold into mountain ranges (*below*).

Thus, about 40 million years ago, the Indian plate rammed into the southern edge of Asia. The result: the Alpine-Himalayan mountain chain, a swath of contorted folds and faults extending all the way from Spain to Southeast Asia. And that epic impact has not yet run its course. Each year, the on-driving plates thrust the Himalayas about two inches upward. But since erosion wears away the lofty peaks at the same rate, their net growth is zero.

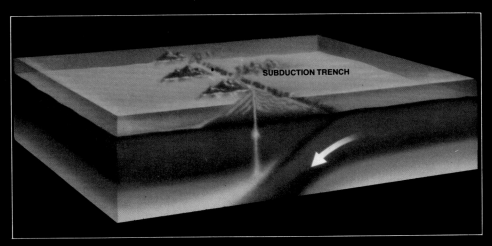

Volcanic island arcs emerge from an ocean where one oceanic plate is descending beneath another. Magma from the partial melting of the descending plate erupts as lava, building land that may eventually become a continent.

In time, two continents are joined in a single land mass, with slices of the overridden plate enfolded in a mountain range that is double the thickness of ordinary continental crust.

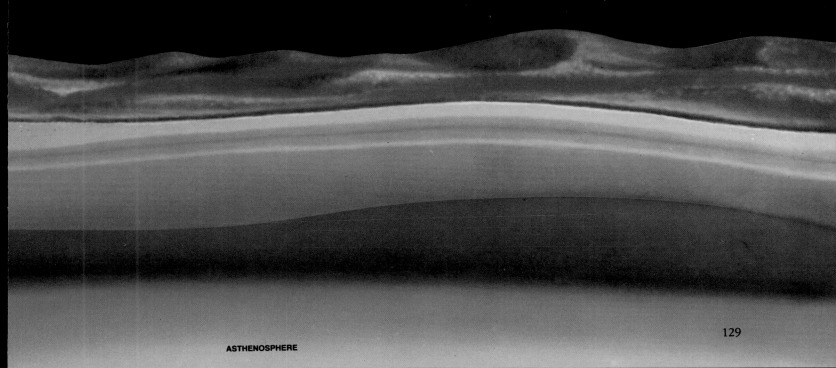

ASTHENOSPHERE

NORTH AMERICAN PLATE

JUAN DE FUCA PLATE

CARIBBEAN PLATE

PACIFIC PLATE

COCOS PLATE

SOUTH AMERICAN PLATE

AFRICAN PLATE

NAZCA PLATE

SCOTIA PLATE

130

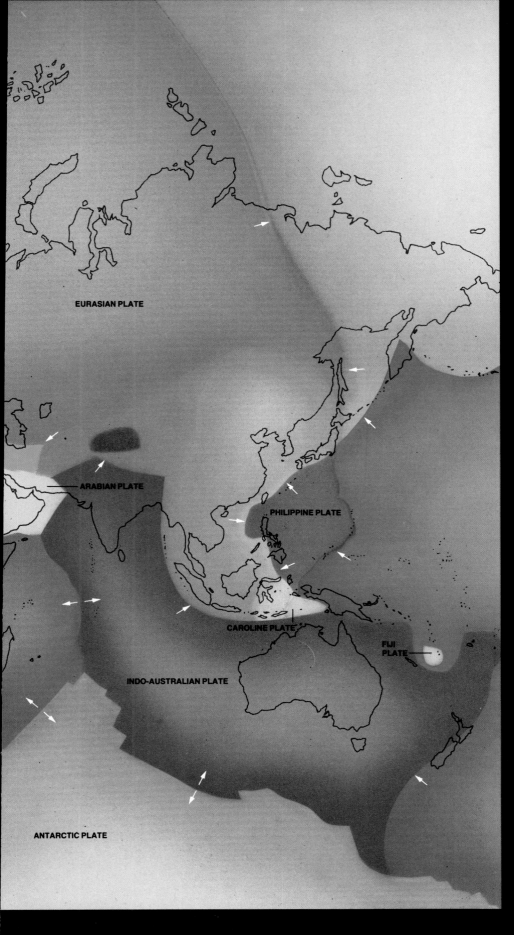

Pieces of Earth: A Puzzle in Motion

A map of the earth's surface based on the tenets of global plate tectonics delineates seven major lithospheric plates and 18 minor ones, including several platelets. Generally, the edges of plates are indicated by the epicenters of earthquakes caused by the jostling and grinding of the moving slabs.

On the ocean floor, earthquakes occur only in narrow belts where the relatively thin plates pull apart or slide past each other, and the plate borders can be plotted with precision. But continental plate boundaries, masked by a thick and varied crust, are far more difficult to discern, especially where they are colliding. From the Mediterranean region to Indonesia and down great swaths of the western coasts of North and South America, plate borders are a diffuse swirl of twisted, contorted and seismically active mountains. Along the southern edge of Alaska, microplates originating in the Pacific have been grafted onto the area south of the Brooks Range, widening Alaska and shifting the subduction boundary seaward.

Plate movements can be gauged more exactly than their boundaries. Sophisticated satellite laser ranging techniques calibrate plate motion over intercontinental distances, enabling scientists to predict earthquakes and other geological events with ever increasing accuracy. Such methods have confirmed that plates with large continental loads, such as the African, Eurasian and American Plates, move the slowest, about three quarters of an inch per year; those with little or no continental crust embedded in them—the Pacific, the Nazca and the Cocos, for instance—move as much as five inches per year.

EURASIAN PLATE

ARABIAN PLATE

PHILIPPINE PLATE

CAROLINE PLATE

FIJI PLATE

INDO-AUSTRALIAN PLATE

ANTARCTIC PLATE

The current configuration of the earth's lithospheric plates, including many platelets, is shown on a map of the world. Arrows show the direction in which each plate is moving.

Chapter 5

THE SLOW DANCE OF THE CONTINENTS

Despite the rapidly established utility of plate tectonics as a framework for understanding the features of the earth, a great mystery remains: What force, or combination of forces, could possibly provide the energy needed to move these huge slabs of lithosphere over the face of the globe? Studies of gravity, magnetism, heat flow and seismic waves have helped build a rudimentary picture of what may be going on in the interior of the planet, but the difficulties of gaining a more detailed understanding are immense. No drill can penetrate a tectonic plate to discover what lies below it; the deepest boreholes in the world sample only the top 10 per cent of the crust. As one researcher has observed, using such limited data as seismic waves to determine the structure of the earth's interior is like trying to figure out the inner workings of a piano by listening to the noise it makes while falling down a flight of stairs.

Given the basic uncertainty, it is not surprising that some of the explanations advanced for plate movement—wobbles in the earth's rotation, gravitational drag exerted by the moon, or a collision with a giant meteorite—have bordered on the fanciful. While most geophysicists will not absolutely rule out the possibility of undiscovered exterior forces, most believe that the vast energy needed to shift continents and crumple mountains must be generated in the interior of the earth, probably by some kind of thermal convection. Richard O'Connell of Harvard University was perhaps as definitive as any earth scientist would care to be when he described the driving force of plate tectonics as "a heat engine that raises hot material to a cool surface, where it eventually sinks and does some work in terms of moving plates around."

Thermal convection has been a familiar scientific concept since its elucidation in the 1790s by the physicist Benjamin Thompson. The process is clearly evident—to pick an everyday example—when a pan of water is put over a flame: Heat causes the water at the bottom to expand and become less dense; as a result, it rises and spreads out on the surface, where it cools. Cooling increases its density again, and the water sinks back to the bottom—there to begin the cycle anew. Such rising and falling patterns of circulation, called convection cells, can be observed on a grand scale in the deep-ocean currents and in the earth's atmosphere.

The possibility that such convection might also take place in the interior of the planet was first proposed at a time when most scientists believed that the mantle was liquid. William Hopkins of Cambridge University advanced this idea in 1839, and in 1881 the British clergyman Osmond Fish-

Under a brightly capped frame that marks his position, a member of a surveying team prepares to take readings in a rugged stretch of the Karakoram Range in Pakistan. When compared with a similar survey made in 1913, the 1980 readings revealed the dimensions of the intervening changes in the terrain, which is still being forged by the impact of the Indo-Australian and Eurasian Plates.

133

er, author of a textbook titled *Physics of the Earth's Crust,* suggested that convection currents under the crust might contribute to mountain building. But these ideas were discarded at the end of the century, after scientists concluded, on the basis of newly available seismic data, that the mantle was solid. Understandably, no one could accept the notion that convection could occur in solid rock. Thus when British geologist Arthur Holmes suggested in the early 1930s—long before the concept of continental drift gained credence—that convection might be the force that split continents apart and then moved them across the face of earth, his idea was resoundingly rebuffed.

When sea-floor spreading was confirmed and the theory of plate tectonics took hold in the 1960s, geophysicists and geologists were obliged to reconsider the evidence for mantle convection. By then, research in fluid dynamics and rock mechanics had indicated that solid rock is not as rigid as had been thought; experiments showed that, when subjected to enough heat and pressure, even the hardest rock will soften until it is capable of a slow deformation called creep. Geophysicists concluded that mantle rock begins to creep at temperatures exceeding 1,835° F. At this temperature the mantle is thought to behave rather like Silly Putty: The silicone-based play material will fracture like a solid if it is subjected to a sudden shock, but if it is left alone, it will gradually flatten into a puddle under its own weight.

The rate of mantle flow is so slow that it would be imperceptible to the human senses: The hour hand of a clock creeps along too slowly for the eye to discern, yet it moves about 10,000 times faster than the apparent pace of mantle material. To geophysicists, however, such a speed is significant because a particle of rock would take a mere 58 million years to move from the bottom of the mantle to the top (the estimated age of the earth is almost 80 times greater). Thus, in geologic terms, the mantle is virtually churning.

Two theories of mantle convection have been put forth. One holds that convection cells are limited to the top 435 miles of mantle rock. The other proposes that the cells circulate throughout the mantle's entire depth of 1,800 miles.

The upper-mantle convection posited by the first theory is supported by a good deal of evidence. Almost all earthquakes, for example, are triggered by plate movement. Most earthquakes occur within the cool, strong plates; many take place in the slabs of lithosphere that are sinking into the subduction zones; but none are recorded at a depth of more than about 435 miles. Analysis of seismic waves indicates that there is a boundary of some kind at that depth. This boundary could be the result of a chemical change in the mantle; if a different, heavier material existed below the 435-mile mark, lighter rocks in the upper area would be prevented by their own buoyancy from sinking more deeply into the mantle.

More support for the theory comes from calculations—also based on seismic-wave analysis—of the stresses exerted on plates as they sink into the subduction zones. Down to a depth of about 180 miles, a sinking slab is often in tension: The forces exerted on it are dragging it down and tending to stretch it. But below 180 miles the slab meets more resistance in the mantle, and tension changes to compression, which is greater in a slab that penetrates more deeply. Presumably, the deeper parts of the slab are encountering increasing resistance to their continued subduction. And since the plates are part of the convective circulation in the mantle, their behavior while sinking strongly suggests that convection is confined to the upper mantle.

On the other hand, there is some evidence to support the alternate notion of whole-mantle convection. In the late 1970s, Harvard geology professor Richard O'Connell, along with graduate student Bradford H. Hager, devised some mathematical models of whole-mantle convection based on the principles of atmospheric circulation. The models suggested that convective flow in the mantle involves a global exchange of material in the same way that air masses gradually mix with one another at their boundaries. The Harvard team assumed that the moving mantle material is not confined to a closed cell under a particular plate but in fact leaks from one cell to another. For example, some of the cold rock going down with the Nazca Plate, which dives under the western coast of South America, might not move back to the East Pacific Rise with the majority of the return flow, but could slip under the South American Plate and flow toward the Mid-Atlantic Ridge. Thus, the convection currents could pick up leaked material wherever plates are being subducted.

To examine the implications of their findings, O'Connell and Hager used the models to predict mathematically the different angles at which plates would descend into the earth in the event of both upper-mantle convection and whole-mantle convection. They found that actual plate orientations, determined by earthquake locations, often corresponded closely to the predictions that assumed flow in the whole mantle, but bore no resemblance to the model of upper-mantle convection cells.

Further support for whole-mantle convection emerged when Thomas Jordan of the Scripps Institution compiled evidence that subducting plates plunge much deeper than 435 miles, and concluded that convection cannot be confined to the mantle above that level. Jordan studied the seismic waves from earthquakes occurring at the very bottoms of the seismic zones beneath two ocean trenches, one under the Sea of Okhotsk north of Japan and the other beneath the Peru-Brazil border. From this data, he sketched three-dimensional pictures of the descending plates. "There are zones of high seismic velocity beneath the deepest earthquakes that look like the continuation of the slab into the lower mantle," Jordan reported. He contended that the only way to explain the data was to assume that the slab continues downward to a depth of between 560 and 620 miles. "Below that depth we lose resolution," he wrote. "My suspicion is that these slabs go very deeply into the mantle, but I don't know how deeply."

One objection to upper-mantle convection—and to convection in general—is based on the geometry of the cells. Studies made during the early part of the 20th Century showed that thermal convection is generally characterized by a symmetrical pattern of cells of similar width and depth. But the plates that make up the earth's lithosphere vary enormously in size, and their widths are usually much greater than the 435-mile depth of the upper mantle. Perhaps, as some scientists have suggested, the upper mantle contains many small, symmetrical cells that churn through the mantle and move the plates. However, if the plates mirror the geometry of convection cells in the mantle, the cells would have to be uniquely flat and asymmetric.

Another problem is that small-scale laboratory experiments with clay and putty indicate that convection in viscous materials does not occur in large cells with long, continuous edges like those evidenced by the mid-ocean ridges. Instead, evenly spaced blobs bubble up through the surface on the top of rising thermal columns. Such columns are familiar to earth scientists, and may offer a quite different explanation of what drives the tectonic plates.

In 1963, a few years before the plate tectonics revolution, Tuzo Wilson of the University of Toronto pointed to the Hawaiian island chain as dramatic proof of plate movement. Stretching to the west and north from the misty plateaus and craggy ocean cliffs of the big island of Hawaii is a string of smaller islands and submerged volcanoes, or seamounts, 3,700 miles long. And every one of these islands and seamounts is thought to have formed in the place where Hawaii now stands.

Wilson suggested that under the Pacific Plate, deep in the earth's mantle directly below Hawaii, there is a "hot spot" that causes a massive thermal plume to rise up and melt holes through the plate like a blowtorch. This subterranean blast furnace is nearly stationary, apparently unaffected by mantle convection, perhaps because of its great depth in the interior of the earth. The Pacific Plate slides over it at the rate of about five inches every year, and when a hole is melted through the plate a new volcanic island appears.

Hawaii itself consists of five coalescing volcanic mountains that were built by lava rising from the mantle. And as the constant rumblings of Kilauea, the world's largest active volcano, attest, the island has yet to move completely away from the hot spot. The farther the other islands in the chain are from Hawaii, the greater their age. Oahu, about 150 miles to the northwest, burst out of the sea about 3.5 million years ago. Midway, one of the oldest islands in the chain, was formed between 15 and 25 million years ago. About 2,000 miles from Hawaii, the chain abruptly veers—suggesting that the Pacific Plate changed course about 40 million years ago—and extends north as a line of submerged volcanoes known as the Emperor Seamounts. Where the chain's long march ends at the junction of the Aleutian and Kurile Trenches off the coast of the Soviet Union, the volcanoes are more than 70 million years old.

A visitor to the Hawaiian chain can see dramatic evidence of its aging. The island of Hawaii is smooth and stony, its volcanic topography virtually intact. The broad beaches and tremendous cliffs and canyons of Oahu and Kauai clearly show that these two older islands have undergone a great deal of erosion from wind and waves. Midway, older still, has been worn away and has sunk beneath the sea; it remains as an island only because of the continual growth of coral on its top.

Similar trails of conveyor-belt volcanoes are found elsewhere in the world. And if the volcanoes happen to migrate from hot spots under the Mid-Ocean Ridge, they are built on each of the spreading plates, and eventually form a V-shaped pattern as the ocean floors move farther away from the ridge. Strung out east and west of Tristan da Cunha in the South Atlantic are two lines of progessively older volcanoes marking, like milestones, the spreading and northward drift of the Atlantic during the last 110 million years. The Galápagos Islands and Easter and Pitcairn Islands in

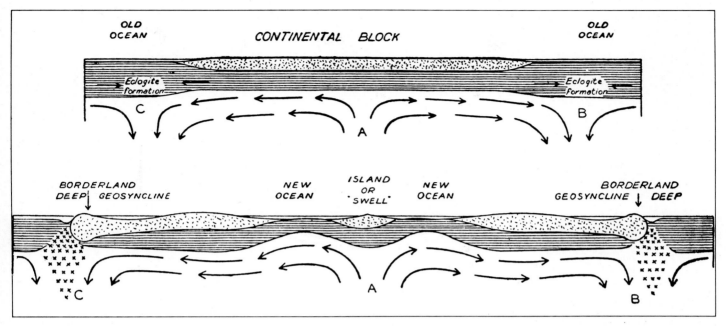

Two 1928 diagrams by British geologist Arthur Holmes depict the convection currents that he believed accounted for continental drift. Material in the earth's semifluid mantle wells up from the hot interior, moves continents apart, and then, as the heat dissipates, descends back into the interior.

the Pacific are parts of other such chains, and a number of island chains in the Indian Ocean also appear to have been caused by hot spots beneath oceanic ridges.

Scientific interest in the hot spots was aroused because they offered convincing evidence of sea-floor spreading. Then, in 1971, W. Jason Morgan, the Princeton geophysicist who had worked out the mathematical description of plate movement, proposed that the huge thermal plumes rising from hot spots might provide enough energy to drive the plates. Morgan theorized that thermal plumes, measuring as much as 60 miles across, carry several trillion tons of hot, viscous rock up from the deep mantle each year, but that only a very small amount—perhaps less than 1 per cent —of the material could rise to the surface through volcanoes or the rift in the Mid-Ocean Ridge. He argued that most of the plume material must spread out under the plates. The flow of viscous rock, he said, would drag the plates along with it.

Other scientists were soon suggesting additional effects of thermal plumes. John F. Dewey and Kevin C. Burke of the State University of New York at Albany proposed that when a rising plume reaches the bottom of the plate and heats it, the plume first manifests itself as a bulging dome very much like a piecrust rising in an oven. If the heating and pressure continue, the dome splits into three cracks, which radiate from the center at angles of about 120 degrees. Many such cracked domes have been discovered, particularly in Africa; they include the spectacular Ahaggar massif in the central Sahara, a great bulge about 1.4 miles high and nearly 60 miles in diameter, and the Afar Triangle, the hellishly hot desert trough in Ethiopia at the junction of the Red Sea, the Gulf of Aden and Africa's Great Rift Valley. Dewey and Burke also suggested that this phenomenon might cause the breakup of continents. If the plumes were close enough together, they said, the cracks in the domes could easily link up to form a great split in the crust.

In addition to his work with Dewey, Burke cooperated with Tuzo Wilson in an attempt to determine the number of hot spots active around the

137

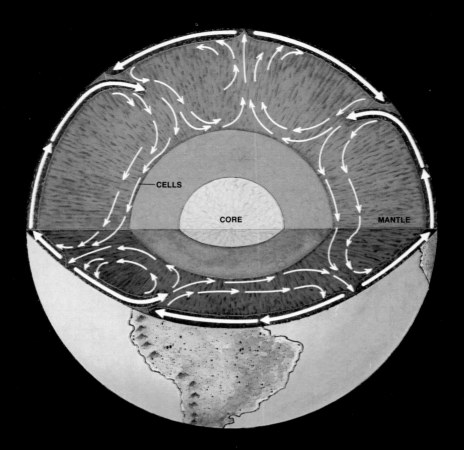

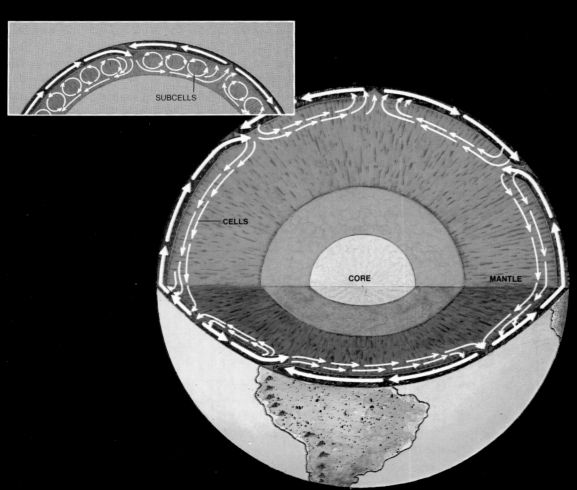

Two possible models of convection cells in the earth's mantle are illustrated at left. Advocates of whole-mantle convection argue that the cells extend about 1,800 miles to the outer core of the earth (top). Opponents counter that the currents descend no farther than about 435 miles (bottom) and perhaps contain rounded subcells (inset).

world during the last 10 million years. Counting every stray volcano or uplift that had not been otherwise explained, they found 122 possibilities—a number that many scientists believe is too high—53 of them in ocean basins and 69 on continents. The greatest concentration was on the African Plate, where they counted 25 on land, 8 at sea and 10 on or near the surrounding midocean ridges. Some of the hot spots on the African continent were found to have lava layers of several different ages superimposed on each other, suggesting that the entire plate is more or less stationary. This led researchers to speculate about a possible connection between the speed of plate movement and the distribution of hot spots. The African Plate covers only 12 per cent of the earth's surface but has 35 per cent of the detectable hot spots and appears to be at a standstill. Other plates with large numbers of hot spots—Antarctica, China and Southeast Asia—are also moving very slowly. By contrast, plates with fewer hot spots, such as North America and South America, are moving more rapidly.

Though differing in detail from convection cells, hot spots are basically a form of convection within the mantle. There is, however, another, wholly dissimilar explanation for plate movement: Perhaps these stupendous slabs are sliding downhill.

At half past five on the afternoon of March 27, 1964, a severe earthquake shook the Alaskan city of Anchorage. Minutes later, a 130-acre chunk of Turnagain Heights, a pleasant suburb overlooking Cook Inlet, slid 700 feet and fell into the sea. Afterward, seismologists discovered to their surprise that the suburb—along with more than 120 million tons of clay, soil and gravel—had slithered down a gentle slope of only 2.5 per cent. This remarkable event encouraged researchers to consider whether the earth's plates are moving in a similar fashion, since many of them slope at about the same angle.

The Mid-Ocean Ridge can be likened to a great welt in the earth's surface, swollen by molten rock that wells up in the rift and keeps the central region hot and expanded. The plates are thus tipped upward at the ridge, and the driving mechanism of plate tectonics could be no more mysterious and no more complex than the force of gravity. Proponents of the idea believe that the sinking slabs might be propelled by trench pull, or what has been dubbed the "washcloth effect." A washcloth will float on water until one edge dips below the surface; then, as more of the cloth sinks and becomes heavier, it descends more quickly, dragging the rest of the cloth under. Something similar might be happening to the earth's plates as they are pulled down into ocean trenches by their heavy leading edges, which have cooled and become denser during their journey across the sea floor.

An early objection to the idea of gravity-driven plates was that some of the plates are so long and thin—the Pacific Plate, for example, is 6,000 miles long and no thicker than 60 miles—that if they were pulled by one end they would simply break up. It was tantamount, said the skeptics, to trying to pull a mile-long noodle by one end. But Walter Elsasser, among the first geophysicists to study plate mechanics, did not agree. He suggested that the rigid lithosphere overlying the softer asthenosphere is so strong that it could easily withstand the stresses of motion caused by gravity.

The washcloth effect was given additional credibility by Frank Press of

Keys to the Earth's Hot Spots

The theory of plate tectonics accounts for most of the world's geologic features in terms of stresses along plate boundaries; but some of the most spectacular formations on earth are produced by so-called hot spots far from the edges of any plate. Although hot spots are volcanic in origin, their scale is far smaller than such tectonic features as the volcanic Ring of Fire along the borders of the Pacific Ocean; instead, they are compact areas of turmoil whose landforms emerge with startling abruptness.

One such anomaly—a series of harrowingly steep volcanic ridges in southern Algeria *(right)*—looms from the sands of the Sahara nearly 1,000 miles from the closest plate border. Another —one of the world's major volcanic islands, Hawaii *(opposite, bottom)*—simmers in the middle of the Pacific Plate.

The characteristics of Hawaii—and of the long sequence of progressively older volcanic islands and seamounts that stretch to its northwest—provide a basic insight into the peculiar nature of hot spots: They appear to be plumes of molten material that well up from a fixed position deep within the earth as the plates drift above them. Thus the same hot spot that formed Hawaii spewed up the older islands in the chain in earlier eruptions.

One theory suggests that the plumes originate amid churning, shallow convection cells that drive the plates; another ascribes them to an even deeper and hotter layer of the asthenosphere. The spouts of magma are so powerful that some experts believe they help fracture old continents and forge new ones.

While the precise causes and long-term consequences of hot spots remain subject to debate, there is no denying their immediate effects. Whether embodied in the fiery guise of a seething lava pool or the icy mask of a thermal-spring terrace *(pages 142-143)*, the earth's deep energy surfaces in hot spots with matchless drama.

Water and wisps of steam escape from the funnel of a geyser at one of Iceland's many thermal springs that are heated by magma. The geologically turbulent island is racked by both hot-spot and plate-rifting activity.

A low sun casts rich light on the basalt cliffs of Algeria's Ahaggar range—relics of a surge of thick lava that crested at this spot six million years ago. Subsequent erosion etched sharp lines in the face of the rock.

Lurid veins of lava steam from yellow fissures in the vent of Hawaii's Kilauea crater. Scientists believe that the molten magma responsible for this 1967 eruption was a product of the hot-spot activity that formed the entire island chain.

The frosty appearance of calcite deposits
belies the presence of scalding mineral water
along the terraces of Mammoth Hot Springs
—progeny of a blistering hot spot that
lurks beneath Yellowstone National Park.

the Massachusetts Institute of Technology, who calculated the densities of the lithosphere and the asthenosphere and concluded that the underlying asthenosphere was actually lighter than the plates riding atop it, a condition that would make it easier for the plates to slide into the mantle under the force of gravity.

At present, gravity drive seems to be the front-runner among the candidates for the motive force of plate tectonics. However, science may never be able to definitively answer this ultimate question about the dynamics of the earth. As Donald Forsyth of Brown University has observed, all assumptions about the driving mechanism are "completely unrestrained by observations of what is going on underneath the plates."

Whatever the nature of the titanic forces that are rearranging the earth's crust, there can be no question that tectonic processes have had an immense global impact—not only on the location of continents, oceans and mountain ranges, but on the distribution and survival of life on the planet, and on the natural resources that have become so vital to modern human society.

Two hundred million years ago, there was just one major continent—dubbed Pangaea by Alfred Wegener—on the face of the earth. Pangaea was a vast terrain of deserts, salt lakes, mountain ranges, tropical swamps, forests and scrubland, roamed by primitive reptiles—among them the first dinosaurs. Along its ragged coastline broke the waves of a universal ocean, Panthalassa, aswarm with marine organisms.

It is probable that Pangaea was not the first such supercontinent. Paleomagnetic readings, geological evidence and fossil distribution suggest that the plates may have been on the move for at least half the earth's 4.5-billion-year history, and plate tectonic activity had almost certainly started by about two billion years ago. Between then and the present, continents drifting at an average rate of 2.5 inches per year could have circled the globe at least four times, traveling 100,000 miles. During their journeys, two or three predecessors to Pangaea may have formed, and the repeated joining and fracturing may have occurred along many of the same sutures. Certainly there is ample evidence of extraordinary continental wanderings. For example, measurements of ancient orientations of Africa and North America to the magnetic poles indicate that those two continents were 6,000 miles apart 450 million years ago: The Sahara was under the great dome of ice that covered the South Pole, and North America straddled the Equator. Yet 250 million years later, the two continents were locked tightly together as part of Pangaea—and, of course, they would eventually drift thousands of miles apart again.

Ophiolite zones—distinctive rock sequences that contain slices of what appears to be former ocean crust, rich in iron, magnesium and copper—are believed to mark the lines along which continental collisions occurred before the formation of Pangaea. By dating the ophiolites found embedded within many deformed mountain belts—the last vestiges of entire oceans that disappeared as continents approached one another—geologists have constructed a chronology of the opening and closing of former oceans and the slow waltz of the continents.

About 500 million years ago, it seems, a supercontinental predecessor of Pangaea began breaking into four pieces. Two of these great land masses collided into each other some 110 million years later, crumpling

Chemical clouds swirl around Venus, obscuring its surface. Only after radar and space probes penetrated the dense cloud cover could scientists begin to study the geology of the planet.

Layered rock formations, perhaps created by a series of lava flows, pave the surface of Venus in this photograph taken by *Venera 14*.

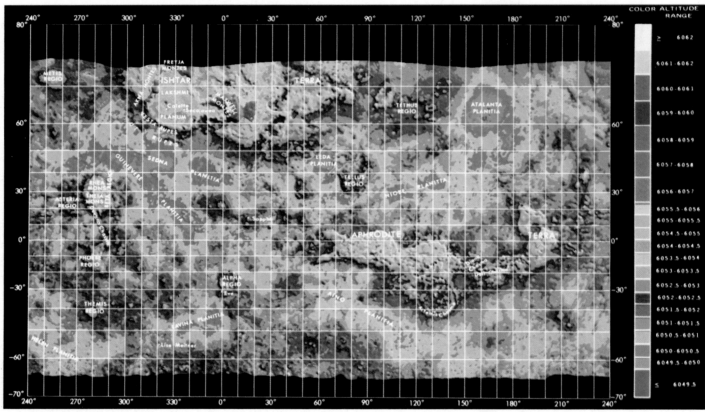

A color-keyed topographic map, based on radar data from the orbiting probe *Pioneer Venus*, reveals the contours of the cloud-veiled surface of Venus. Scientists believe that many of the uplifts (*yellow and orange*) and depressions (*dark blue*) were shaped by volcanic activity.

A rift valley 1,400 miles long and 175 miles wide scars the Venusian landscape. This artist's conception exaggerates the rift's 9,500-foot depth but clearly shows the high-shouldered profile that has led some geologists to speculate that the valley was formed by volcanic uplift, followed by faulting.

Signs of Tectonics on an Alien World

Unmanned Soviet and American space probes visiting Venus, the earth's nearest neighbor in space, have found that although the searing, 850° F. Venusian surface is a biological wasteland, the planet exhibits a kind of geologic life. After parachuting through Venus' atmosphere—a witches' brew of carbon dioxide and wind-whipped clouds of sulfuric acid—the Soviet planetary lander *Venera 9* in 1975 transmitted a crude image of the surface that astonished geologists. Wind-blown dust and the corrosive atmosphere quickly smooth any jagged features on the surface of Venus, but a profusion of rocks clearly visible in the photograph had sharp edges, indicating they had been formed recently, perhaps by a volcanic eruption. Two 1982 landers, *Venera 13* and *Venera 14,* beamed back more evidence of volcanic activity, including photographs showing tracts of rock that closely resemble lava flows on earth.

While Soviet landers investigated Venus at ground level, an American probe, *Pioneer Venus,* began orbiting the planet in 1978, scanning its surface with a radar mapping device. On the relief maps that resulted, the signs of geologic activity were striking: crater-topped peaks, apparently volcanoes; low-lying basins, perhaps floored with lava flows; and a vast rift valley.

Despite the evidence of geologic unrest, many scientists doubt that the horizontal movements of earth-style plate tectonics occur on Venus. Instead, they believe that tectonic activity centers on random volcanic hot spots, which pour out magma until the surrounding crust becomes overloaded and sinks into the mantle, melting and renewing the magma supply. A similar process of vertical tectonic movements, suggest some geologists, may have been the norm on earth three billion years ago, before the lateral motions of plate tectonics began.

like wrinkled rugs along their leading edges and creating the Caledonian mountain system, which today stretches across northwestern Europe and continues as the "old" Appalachian chain in northeastern North America. An Africa-North America crunch folded the more southerly "younger" Appalachians about 300 million years ago, and the collision that welded Siberia onto Pangaea, probably about 250 million years ago, resulted in the squeezing up of the Urals, the 1,600-mile-long mountain range separating Europe and Asia.

Pangaea covered two fifths of the earth's surface. Its irregular coastline was marked most noticeably by a large V-shaped notch about halfway along the continent's eastern edge, an enormous triangular body of water known as the Tethys Sea. North of the Tethys was Laurasia, descendant of an earlier and smaller continent called Laurussia; it consisted of about 30 million square miles of primordial territory, which would eventually split to become North America and Eurasia. South of the Tethys was Gondwana —which the 19th Century Austrian geologist Eduard Suess had named Gondwanaland. Similar in size to Laurasia, it included the future South America, Africa, India, Australia and Antarctica.

The presence of so prominent a land mass in the Southern Hemisphere probably had a major effect on the earth's weather. One theory proposes that ocean circulation—the principal mechanism for transferring heat from the Equator to the Poles and thus moderating the climate—was restricted by Pangaea. A great ice age set in. Pangaea was located so far to the south that from time to time much of Gondwana was covered by a massive ice sheet. The ice encased not only land that would become Antarctica, but most of present-day India and Australia, and the southern parts of Africa and South America as well. Even ice-free regions of Gondwana had very severe climates.

But as Pangaea drifted slowly northward, the icecap began to melt, raising the oceans until they swept from Pole to Pole, triggering not only a global warming trend but a dramatic change in the evolution of life. Before Pangaea had begun to break up, the first ungainly land dwellers— amphibians, reptiles and mammal-like reptiles—had spread widely across the continent. But the overwhelmingly dominant forms of life on earth were marine creatures until, about 225 million years ago, the warming caused by the supercontinent's movement encouraged a worldwide increase in land-based life.

By about 200 million years ago, as much land had moved into the Northern Hemisphere as remained in the Southern. The Y-shaped junction of North America, South America and Africa was situated where Ascension Island can be found in the South Atlantic today. The site of New York was on the Equator, Japan was in the Arctic, while India and Australia were far to the south, bordering Antarctica.

Except for its great mountain chains, the terrain of Laurasia was lower than that of its southern counterpart, Gondwana, and was susceptible to flooding. At different times, great areas of North America, Europe, Siberia and China were covered by shallow seas. Elsewhere the land was predominantly desert or sparsely vegetated scrub and conifer forests. The climate was generally dry and mild, with hot, dust-filled winds eroding the hilltops and filling the valleys with rubble and sand. Rivers were ephemeral, although they could be swollen by storms into torrents of muddy water.

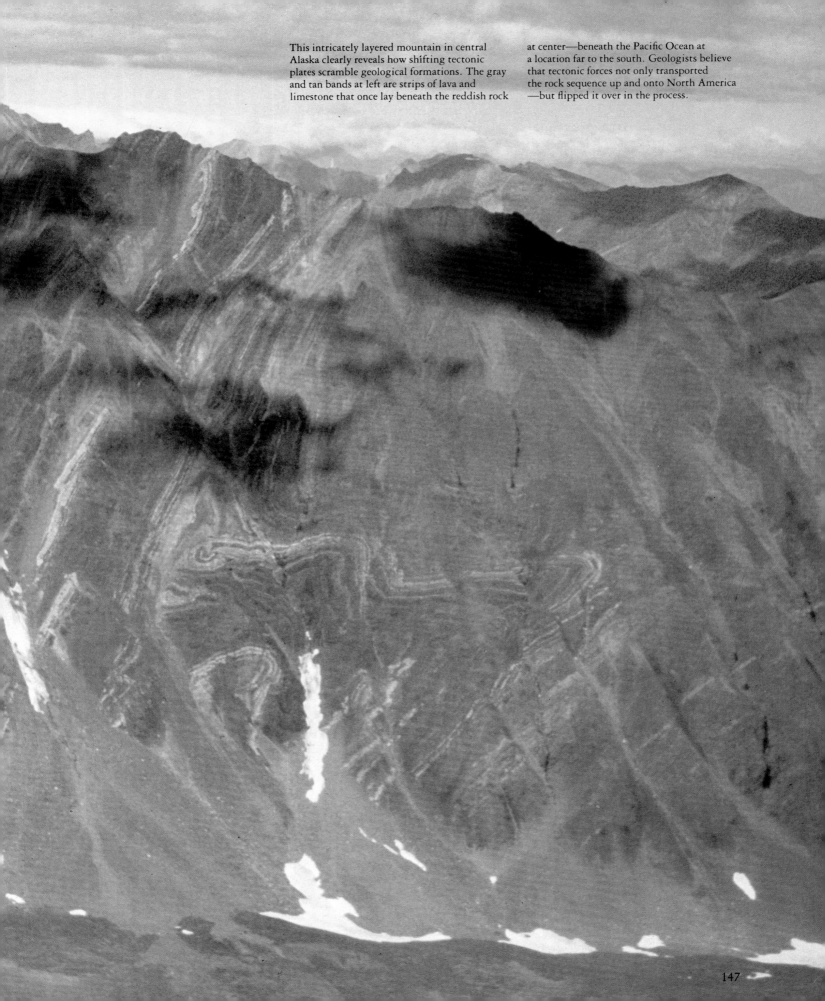

This intricately layered mountain in central Alaska clearly reveals how shifting tectonic plates scramble geological formations. The gray and tan bands at left are strips of lava and limestone that once lay beneath the reddish rock at center—beneath the Pacific Ocean at a location far to the south. Geologists believe that tectonic forces not only transported the rock sequence up and onto North America —but flipped it over in the process.

147

Around the shores of the Tethys Sea were brackish swamps and lagoons choked with decaying vegetation.

Although earth scientists believe that the network of fractures along which Pangaea was to rift existed when the supercontinent was formed, they cannot say for certain what triggered the cataclysmic outpouring of basaltic rock that at last split the continent asunder. What probably happened, though, was that Pangaea's weak links opened little by little, allowing increasing volumes of lava from the earth's mantle to bubble to the surface. The spreading and thinning of the continental crust and the upwelling of this heavy basalt onto the surface caused the area near the rift to fracture anew and sink, forming basins in which water repeatedly collected and evaporated. Eventually, a fracture spread to the coast, and sea water from the Tethys or from Panthalassa flooded into the basin, creating a long narrow sea between embryonic continents. These inlets were the forerunners of today's deep oceans, and the fractures became midocean ridges.

The sweep of time during which these events occurred is so vast as to defy comprehension. It helps to imagine the geological clock running much faster, compressing millions of years into minutes. Scientists disagree about the timing and direction of past tectonic plate movements, and the evidence is frequently ambiguous and contradictory; a reasonable scenario of plate wanderings over the past 200 million years can be constructed from the work of a number of researchers. Suppose, then, that the sprawling continent of Pangaea lies seamless and solid at high noon of a portentous day. In the next hour, the whole prodigious planetary drama of two million centuries will take place.

Seconds into that hour, the breakup begins as a great rift splits Pangaea from east to west just above the Equator. As the fracture first opens, molten rock wells up from the mantle into the widening gap between Laurasia—consisting of North America and Eurasia—and Gondwana. The weight of the solidified magma causes portions of the crust to fracture and drop, until the waters of Panthalassa pour in to create a narrow body of water (similar to the Red Sea, which today separates the Arabian Peninsula from Africa). This rift almost completely separates the two great land masses of Laurasia and Gondwana: They remain in contact only where the southern rim of Spain touches the North African coast near Morocco. This contact acts as a "hinge" as the Laurasian plate begins rotating clockwise, slowly widening the rift in the west (opening the Atlantic) and narrowing the triangular wedge of the Tethys Sea in the east (closing the Mediterranean). Meanwhile, in the south, Gondwana begins to split in two, with South America and Africa on one side of the rift, and India, Madagascar, Antarctica and Australia on the other.

Twenty minutes of the hypothetical hour, or about 65 million years, have passed, and the supercontinent Pangaea is no more. The gap between North America and the block made up of South America and Africa is growing steadily wider. East of the Spanish-Moroccan hinge, an ocean trench running from Gibraltar to a point near Borneo consumes the ocean floor overrun by the rotation of the Laurasian plate.

During the next 10 minutes (33 million years), the embryonic North Atlantic Ocean continues to widen until it is more than 600 miles across; the rift snakes northward, wrenching open a fracture between Canada and Greenland. The east coast of the United States is now aligned east-west

When the German research vessel *Sonne* (sun) left Western Samoa in June of 1981, its destination was a closely held secret. The scientists aboard were in search of treasure—in the form of dark, mineral-rich deposits on the ocean floor.

In particular, they were after cobalt, a rare element used in a number of basic industries, including the production of heat-resistant steel. Cobalt was already being mined profitably in the central Pacific basin, even though it came embedded in other minerals in concentrations as low as .2 per cent. A few studies had stated that cobalt in concentrations of 2 per cent could be found on the slopes of submarine mountains well away from the customary sources: The German scientists intended to test those accounts.

There was no theoretical reason to expect that they were true. The minerals the *Sonne* was seeking are usually spewed into the sea by springs at un-derwater tectonic ridges and travel in plumes of fine particles for thousands of miles before precipitating over vast areas of the seabed. There, in one of the slowest chemical reactions known —spanning millions of years—the precipitates collect in thin, brittle crusts or grow as nodules around shards of rock or fishbone.

During their 30-day voyage, the crew of the *Sonne* used radar scans, cameras and scoops *(below)* to survey and sample steep underwater terrain. Aided by a cable-operated dredge, they retrieved a satisfying quantity of sizable nodules. And later analysis showed that the mineral crusts scraped up from the slopes did indeed contain cobalt in concentrations as high as 2 per cent. By confirming the presence and accessibility of these rich deposits, the *Sonne* had found treasure indeed—assurance of an inexhaustible supply of cobalt for world industry.

Aboard the *Sonne* in the mid-Pacific, a crew member props a radar chart of an underwater slope against a bulbous scoop used to gather mineral samples from the sea floor.

Sailing under a spectacular Pacific sky,
the ungainly *Sonne* boasts an imposing array
of antennas and equipment; the gaping
arch astern supports the heavy sea-floor dredge.

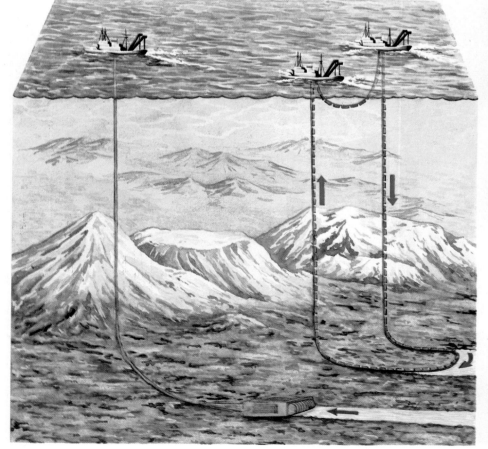

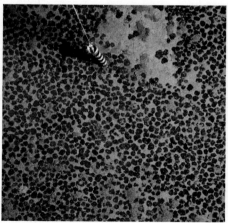

A television camera linked by cable to a monitor
aboard the *Sonne* descends toward a batch
of sea-floor mineral nodules. The striped sinker
drops off on impact with the bottom,
leaving the camera positioned a few yards above
the ocean floor to survey the prospects.

Minerals can be mined from the sea floor with
a chain of dredging buckets that passes
between two ships *(left)*. But a more promising
method *(far left)* makes use of a powerful
"vacuum cleaner" that traverses the seabed and
pumps nodules to the vessel above.

A crew member signals thumbs up to the
Sonne's most recent catch, which includes black,
cobalt-rich chunks amid the light-colored,
chalky sediments *(above)*. The dredging cart,
which is screened at the sides to allow
mud and water to escape, is raised and lowered
with a block and tackle attached to the
ship's sturdy mobile arch *(left)*.

Resembling rings on a tree stump, sections of
nodules reflect patterns of mineral precipitation
on the sea floor. While a tree ring represents
a year's growth, a few millimeters of a nodule
can take a million years to form.

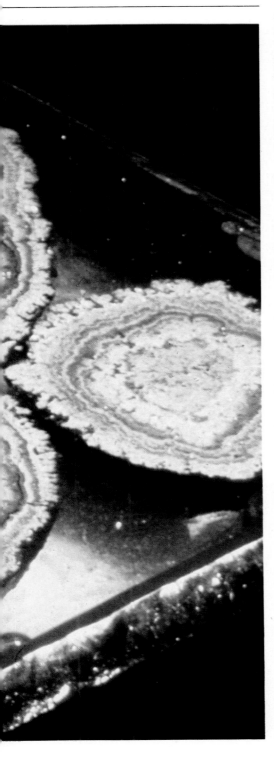

along a northerly latitude of 25°, and coral reefs are growing all along the edge of the Atlantic continental shelf. The jostling of the Eurasian and African Plates forces the Iberian Peninsula to rotate counterclockwise 35 degrees, opening up the Bay of Biscay.

Then, at 12:30, about 100 million years after Pangaea first split, a fissure appears between Africa and South America, extending from the southernmost tip of both continents all the way to Nigeria. This fracture marks the birth of the southern half of the Mid-Atlantic Ridge and the opening of the South Atlantic.

During the next 35 million years, between 12:30 and 12:40 on the hypothetical clock, the world as we know it begins to take recognizable shape. The rift between Africa and South America is completed, and the Mid-Atlantic Ridge becomes a continuous underwater mountain range separating the two continents. At its northern end, the rift migrates from the west side of Greenland to the east, partially splitting Greenland from Europe. Farther to the north, however, Eurasia and North America remain in contact, for the time being.

Africa, freed at last from the embrace of South America, swings like an enormous pendulum north and east, breaking contact with Europe at the Spanish-Moroccan hinge, but almost closing the Tethys Sea at its eastern end. At the same time, North America and South America—separated by thousands of miles of ocean—are drifting steadily westward, colliding on the way with volcanic island chains and stray blocks of oceanic and continental crust that crumple the coastal margins to form the Rocky Mountains of North America and the Andes of South America.

Meanwhile, New Zealand breaks away from Australia and Antarctica, and not long afterward Australia and Antarctica are finally wrested apart. The Mid-Atlantic Ridge has now reached the Arctic Sea, splitting Greenland from Europe. Virtually all that remains to complete the present-day map is for India—which has broken off from Antarctica—to finish its long northward trek across the Tethys Sea. Along the way, it passes over an active hot spot; a giant thermal plume forges through the Indian plate, pouring out a flood of molten basalt that hardens to form the 200,000-square-mile Deccan Plateau.

Forty-six minutes into the hypothetical hour, or some 45 million years ago, India crosses the Equator. A little later, the Indian plate, pushing in front of it ancient sediments laid down millions of years earlier in the warm waters of the Tethys, rams into Tibet with such force that the plate boundaries buckle and fold into great crustal waves, thrusting the mighty Himalayas five miles into the sky and raising the world's tallest mountain peak, Mount Everest. Fossils of marine life that had existed in the time of Pangaea are gathered up in the turmoil and stranded miles inland, high on the craggy mountain peaks. After the first collision, the Indian plate, still pressing north, plows under Asia, raising the Tibetan plateau and fracturing and deforming much of southeastern Asia.

Farther to the west, the African tectonic plate is grinding against Eurasia, slowly squeezing the Mediterranean and buckling a contorted mountain chain along the boundaries of the plates from the Atlas Mountains in northwest Africa across the Alps and through the Caucasus to the Himalayas. Italy, originally part of Africa, barges into Europe and thrusts the Alps still higher.

With seven minutes left in the hour (24 million years ago), water flowing down the eastern slopes of the steadily rising Andes in South America carves the Amazon's 3,900-mile course across Brazil to the Atlantic. At about the same time, the Gulf of Aden and the Red Sea open when a spur of the Indian Ocean ridge splits Arabia from Africa. Shortly afterward, Africa and Asia are welded together in the Middle East, enclosing the eastern end of the Mediterranean.

Another, and rather different, geological extravaganza begins at 12:56 (15 million years ago). As Spain is squeezed against Africa, a shallow shelf of rock across the Strait of Gibraltar rises up until it separates the Mediterranean from the Atlantic. With only a few rivers sustaining it, the Mediterranean begins to dry up. It is quickly reduced to a series of brine pools and lakes, which, as they repeatedly evaporate and refill, create salt beds thousands of feet thick. Soon the Mediterranean is a dry valley, two miles deep. Then, at 12:58, or about six million years ago, the mountain wall holding back the Atlantic is breached, and the ocean begins to cascade over the rock shelf back into the Mediterranean. About 5 per cent of the world's sea water thunders over the Gibraltar Sill in a mammoth waterfall that lasts about a century—or about $\frac{1}{500}$ of a second on the hypothetical clock. By the time the Mediterranean is full again, water has flooded 750 miles up the Nile valley and the level of the world's oceans has dropped by about 40 feet.

During the last minute of this hypothetical hour, two more events of significance take place. At 12:59, or about 3.5 million years ago, volcanic activity caused by plate movements under the Pacific builds the Isthmus of Panama up out of the sea, joining North and South America and completing the familiar picture of today's world. And in the last seconds before the end of this hour of tectonics, the first members of the genus Homo appear on the planet.

All along, of course, a great deal more has been going on than mere alterations in geography. Intertwined with the grand progress of the continents across their watery parade ground has been the evolution and distribution of life itself.

Biologists have long understood that living creatures exist in a complex and delicate balance with one another and their surroundings. As conditions change over long periods of time, life forms change, too, evolving new characteristics that enable them to survive in a different environment. The theory of plate tectonics introduced a new element into the history of evolution—an entirely different concept of the global changes that affected early adaptations of life on land and in the sea.

While Pangaea existed, the absence of ocean barriers permitted animal species to roam freely across virtually the entire surface of the supercontinent. They were thus able to avoid territory where the climate was inhospitable or where food was unavailable. With no need to adapt to different environments, and with competition among the intermingled species limiting survival to the most successful animals, only a very small number of groups emerged. In the same fashion, Pangaea's enormously long coastline—a continuous shallow-water margin with no major physical barriers and little variation in temperature—did not encourage wide diversification of marine species.

When the breakup began, land animals cast adrift on isolated chunks of the supercontinent found their world circumscribed and escape from it impossible. In order to survive, they had to adapt to the changes that overtook their habitat; at the same time, competition was much reduced, and new species had a better chance of surviving. As a result, many new land-animal orders evolved, most of which are still extant. A similar pattern was followed on the continental margins. The sundering of Pangaea created much more coastline, offering new and different marine habitats, and an unprecedented diversity of specialized marine creatures emerged to fill these ecological niches.

Plate tectonics has helped eliminate species as well as create new ones. Dinosaurs may be a case in point. The multiple collision that formed Pangaea roughly coincided with the beginning of the Age of Reptiles, which was to span nearly 200 million years. Egg-laying reptiles dominated evolution longer than any other form of land animal: The land, the oceans and the skies were all the domain of early crocodiles and lizards, turtles, tortoises and a variety of exotic creatures—such as flying pterosaurs—with no modern counterparts. For much of this time, dinosaurs ruled as the undisputed lords of the earth.

Dinosaurs originated some 200 million years ago, possibly in Gondwana, and they quickly spread over the whole of Pangaea. Some walked on two legs, some lumbered along on four. Some dinosaurs could run at considerable speed if danger threatened, some could climb trees. They adapted to life in dry upland areas and lowland swamps, and grew to gigantic sizes. Then, approximately 65 million years ago, the dinosaurs suddenly vanished from the earth.

Scientists seeking an explanation for this disappearance have proposed a number of alternatives, among them a worldwide catastrophe caused by the impact of a giant meteor. But another scenario, suggested by plate tectonics, holds that the northward drift of the continents caused a marked fall in temperature and an increase in seasonal variation. The dinosaurs, unable to escape from the continents on which they were marooned and too big to find shelter, might have succumbed to the cold along with many other types of reptiles.

Only about 20 reptilian orders evolved during the entire 200 million years of the Age of Reptiles. But the mammals, which were to take over the rapidly dispersing continents of the world, diversified much more widely in order to survive. Indeed, in less than a third of the time spanned by the Age of Reptiles, the Age of Mammals managed to produce half again as many orders.

Primitive mammals had first emerged at a time when sufficient land links still existed for these small, shrewlike creatures, scurrying in the undergrowth, to spread themselves across the embryonic continents before rifting cut them off completely. At the time the dinosaurs disappeared, the mammals—with their more versatile anatomy and more vigorous offspring—were well equipped to survive ecological changes on the drifting continents and to seize their evolutionary opportunity.

As old Pangaea slowly tore apart, each separate block of land became the scene of what biologists call adaptive radiation—an evolutionary process in which, as a single species adapts to different environments, descendant forms diverge more and more from one another. By developing the ability to live in deserts and swamps, tropical forests and barren mountains, by learning to swim, fly, climb, run, eat plants or eat flesh, the mammals quickly filled the ecological gap left by the reptiles and accelerated their own evolution.

In Laurasia, 16 mammalian orders emerged and scattered widely through North America and Eurasia, crossing by way of Greenland, where the two continents were still connected. Moles, hedgehogs and shrews were among the first to make an appearance in Laurasia, along with such early primates as monkeys and apes. Rodents—rats, mice, squirrels and beavers—arrived later, between 55 million and 65 million years ago, as did such carnivores as cats, dogs, bears and weasels. Grass appeared, and along with it came the grass eaters—horses, cattle, deer, pigs and other hoofed animals.

In Gondwana, where rifting had separated the continents much more than it had in the north, markedly different mammals evolved. South America's fauna included ant bears, sloths, armadillos, a species of now-extinct hoofed herbivores, and small marsupials—animals that are born while still in a larval stage and that are nurtured in the mother's pouch. Africa had mastodons and elephants, conies, aardvarks, and aquatic dugongs and manatees. Little is known of the early mammals in India and Antarctica. But it is clear that the placental mammals—which were rapidly dominating evolution, largely because their young are born in a relatively

A diagram of the areas in which the fossils of three long-extinct reptiles have been found reveals narrow, sharply defined habitats extending across three continents and the subcontinents of Madagascar and India. The shape of the animals' ranges can best be explained by assuming that these lands were once united as outlined below.

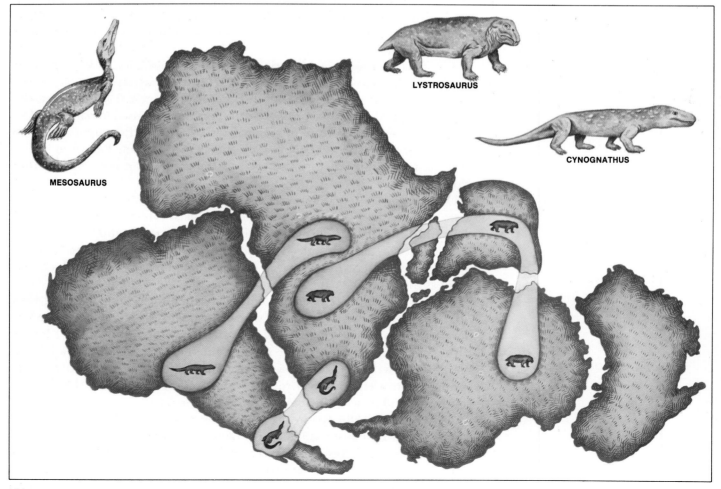

MESOSAURUS

LYSTROSAURUS

CYNOGNATHUS

A jagged jawbone with several teeth intact (*above*)—the first fossil of a land mammal ever recovered in Antarctica—is a relic of the rodent-like marsupial represented in the drawing at right. Found in March of 1982, the 60-million-year-old fossil proves that Antarctica once had a temperate climate and was joined with South America and Australia, places where marsupials survive today.

advanced stage of development—did not win a foothold in Australia. When that continent broke away from Antarctica and sailed off across the Indian Ocean, it bore only the marsupials that had originated there.

When continents collided, or when they were temporarily linked by some form of land bridge, there was an immediate intermingling of originally different species. The fossil record indicates that a land bridge existed at various times across the narrow Bering Strait, which today separates North America and Asia. This bridge assisted the exchange between North America and northeastern Asia of a wide variety of mammals—tapirs, rhinoceroses, cats, rodents, rabbits, hares and many others. After rifting finally separated North America and Europe, the trans-Bering bridge prevented the total isolation of North America.

Some 45 million years ago, the bridge seems to have been inundated, since the mammals on each continent for a time became increasingly distinctive. When the connection was subsequently reestablished, Africa's mastodons and elephants, which had spread through Europe and Asia after the African Plate welded itself onto Eurasia, reached North America via the Bering crossover. Also making good use of the crossing some 12 million years ago was the ancestral horse, a three-toed animal about the size of a large pony, with long legs suited for running rapidly over hard ground. This herbivore originated in North America and dispersed widely over Eurasia after crossing the Bering bridge. It would fare better in its adoptive lands than it did in America, where it eventually became extinct. Descendants of this early horse returned to America with the Spanish conquistadors, who took horses with them in the 16th Century when they set out to conquer the New World; inevitably, some escaped, and soon herds of wild horses were thriving on the western plains, turning the native Indians into nomadic horsemen.

When the Isthmus of Panama at last linked the two Americas more than three million years ago, competing species battled desperately for survival. Only a few South American mammals, notably armadillos, ground sloths and porcupines, managed to establish themselves successfully in

the north. But a procession of mastodons, horses, tapirs, peccaries, llamas, deer, large and small cats, weasels, raccoons, bears, wild dogs, squirrels, mice, rabbits and insectivores flooded into South America—with disastrous effects on the native population. Before the linking of the continents, 29 families of mammals inhabited the south and 27 entirely different families lived in the north. After the union, the two land masses had 22 animal families in common, and most of them were originally from the north.

Among the South American species wiped out were all but two of the many marsupials that had evolved there. Wherever marsupials faced aggressive competition from the more intelligent and adaptable placental mammals, they were invariably the losers. Only one haven remained for the luckless marsupials: Australia. In splendid isolation from the moment it rifted from Antarctica, Australia was the sole continent to retain its pristine

Docile marsupials—here represented by a koala, a kangaroo and an opossum—dominate Australia's ecology. Because the continent drifted into isolation some 50 million years ago, they did not have to contend with the more competitive mammals that developed elsewhere, and nearly 200 marsupial species flourish in Australia today.

marsupial population—a diverse collection of kangaroos, wallabies, wombats, bandicoots, koalas, opossums and phalangers.

Perhaps the most surprising evolutionary effect of the breakup of Pangaea is its continuing effect on certain species. For example, a type of green sea turtle living on the coast of Brazil puzzled biologists for years with its compulsion to swim nearly halfway across the Atlantic to breed on the island of Ascension. Why the turtles would undertake such a long and hazardous journey remained a mystery until the theory of plate tectonics provided the answer: This behavior pattern was probably programed into the turtles' genes 100 million years ago, when a rift slowly opened between South America and Africa. Turtles often prefer to nest on offshore islands to protect their eggs from predators, and it is likely that the ancestors of these sea turtles used a volcanic island formed by one of the hot spots that caused the rift. As the ocean gradually widened, a succession of islands would have risen at the rift, moving away with the downward-sloping sea floor and eventually sliding under the waves. When the first island disappeared, the turtles, guided by the sun, kept swimming until they reached the next one in the chain. Each successive island tempted the determined creatures to swim greater and greater distances, as their home on the Brazilian coast moved farther from their breeding ground, until today their offshore refuge is more than 1,000 miles away.

The insights supplied by the theory of plate tectonics are by no means limited to the realm of pure knowledge. Fitting together the tectonic-plate jigsaw puzzle has many practical applications for a world increasingly in need of raw materials and energy resources—among them, oil.

When the continents tore apart tens of millions of years ago, their rifting created long ocean basins in which deep layers of oil-forming sediments—and, on occasion, salt—accumulated. The resulting petroleum deposits were themselves split by further sea-floor spreading and are now trapped beneath thick layers of rock salt under the continental shelves. Thus when oil is discovered on one side of a present-day ocean, a petroleum geologist can use his knowledge of how the continents once fit together to predict where oil can be found on the opposite side. Much of this offshore oil lies too deep for economical drilling. But if refinement of the theory of plate tectonics makes it possible to pinpoint the location of an oil field before a single hole has been drilled, it may fundamentally alter the economics of oil exploration (currently, an average of nine dry holes are sunk for every strike).

The formation, shifting and jostling of the earth's plates have also been responsible for the creation of exploitable deposits of various mineral ores. While the exact nature of the process is unknown, many researchers believe that some types of ore are probably formed by an interaction between sea water and the upwelling basalts at oceanic ridges where plates are rifting apart. Such activity has been detected beneath the Red Sea, where the African and Arabian Plates are splitting, and where brine pools rich in copper, gold, silver, iron, nickel, zinc and cadmium are forming in the rift. Similar refining processes have been observed by divers exploring the midocean ridges in the Atlantic and especially in the Pacific, where the rapid spreading rate supplies large amounts of heat.

This undersea formation of mineral deposits can explain why so many

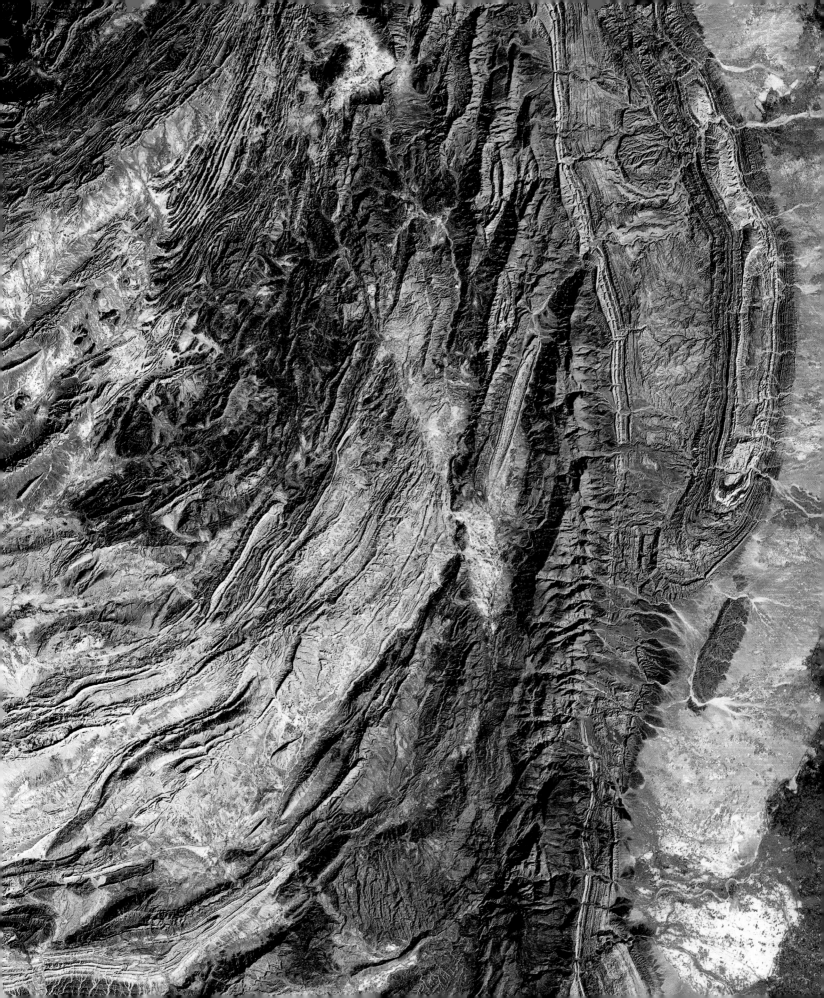

major ore bodies have been found locked in continental mountain ranges
that were formed over a descending oceanic plate or by a volcanic island
arc's collision with a continent. It appears that, as the slowly moving sea
floors with their mineral cargoes finally arrive at subduction zones and dive
into the mantle, the minerals are further refined and then percolated upward. If the oceanic plate is diving beneath a continent, mineral lodes will
be emplaced in the resultant mountain chain. If the plate is sinking into an oceanic trench, the metal-rich solutions will rise in the same fashion to the volcanic island arc forming above; later, as the islands crush
into a continent, the minerals will be entombed in the coastal mountains
built up by the impact.

Moving ceaselessly over the surface of the globe, propelled by forces generated deep within the seething mantle of the earth, the tectonic plates have
shaped the world as we know it, and they will shape tomorrow's world.
Mineral riches and mountain ranges, deserts and seas and the living creatures of the land, water and air—all have been formed or influenced by the
shiftings of these slabs of the earth's crust.

The detection and comprehension of these shiftings is a scientific feat
every bit as remarkable as the discovery, four centuries ago, that the earth
moves around the sun. As with that earlier revolution, the truth was wrung
from monumental, painstakingly assembled volumes of data. And it was
won through intellectual courage of the highest order. Copernicus, whose
lifework challenged the pervasive 16th Century belief that the earth was the
center of the universe, would have found much to admire in Alfred Wegener, the modern-day German genius who suffered so many years of scorn
for his inspired vision of drifting continents and widening seas. Ω

THE ONCE AND FUTURE EARTH

The history of the earth's crust over the past half billion years is largely the story of the assembly and breakup of the supercontinent Pangaea, whose name means "all lands." By studying traces of magnetism preserved in rock, which indicate a continent's orientation and latitude relative to the North Magnetic Pole at the time of the rock's formation, and by examining fossil deposits, which reveal the climatic conditions of the distant past, scientists have plotted continental positions as far back as 550 million years. At that time, and for the next 250 million years, fragmentary protocontinents—borne on great tectonic plates—wandered the globe, drawing closer together.

The gradual collision of those protocontinents, starting about 300 million years ago, formed Pangaea and thrust up mountain ranges that still exist. Later, beginning 200 million years ago, colossal rents, the precursors of the Atlantic

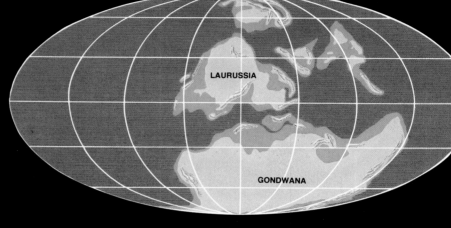

About 320 million years ago, precursors of today's continents converged. To the north lay Siberia, Kazakhstania and China. Laurussia combined Scotland, Greenland, North America, Scandinavia and Russia, while Gondwana made up the rest of earth's land mass.

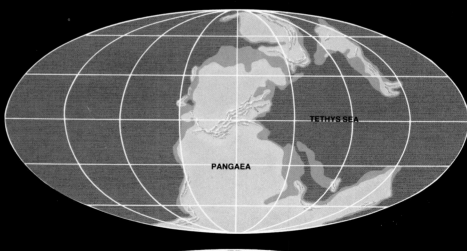

The shape of Pangaea emerged about 250 million years ago. The collision of Laurussia and Gondwana raised a mountain belt, present-day remnants of which are the Appalachians along North America's eastern coast and highlands in central Europe.

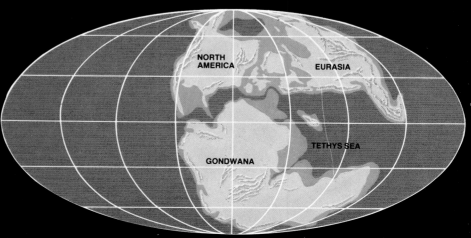

By 135 million years ago, the breakup of Pangaea was well under way. Rifting and sea-floor spreading opened the nascent North Atlantic Ocean between North America and the old continent of Gondwana, in which South America and Africa were still joined.

and Indian Oceans, split Pangaea into the present continents, ending its geologically brief life. This ponderous, inches-per-year dance of continents— one version of which is reconstructed below in computer-plotted maps that portray the entire surface of the earth— continues to reshape the planet today.

Few geologists doubt that the dance will continue, and many believe that it may eventually reunite the continents —although there is no certainty on this point. The pages that follow present a possible scenario of future movements. Developed by University of Chicago geologist Christopher Scotese, with the help of a computer, these projections are based on educated guesses about the past and about the development of new zones of rifting and subduction in ages to come. The scenario's end result *(pages 168-169)* is a future Pangaea, which in unimaginably distant ages may itself break apart and begin the cycle anew.

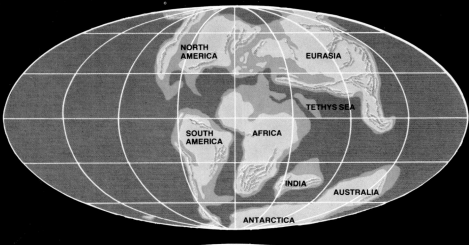

One hundred million years ago, South America and Africa separated, widening the forerunner of the South Atlantic Ocean. Other rifts divided what remained of Gondwana, freeing India to begin its 5,000-mile, 80-million-year journey northward toward Asia.

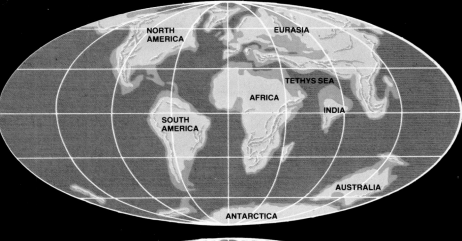

Forty-five million years ago, India crossed the Equator in its rapid move toward Asia. Africa drew closer to Europe, narrowing the ancient Tethys Sea. And Gondwana's breakup was completed when Australia split from Antarctica and drifted northward.

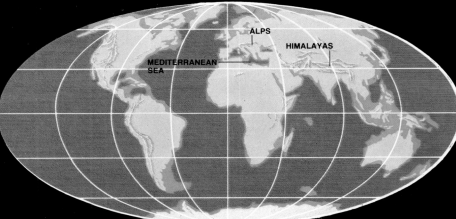

Today, the folded belts of the Himalayas bear witness to India's continuing collision with Asia, while the jostling of the African Plate and the Eurasian Plate still causes the Alps to rise and the last vestige of the Tethys, the Mediterranean Sea, to shrink.

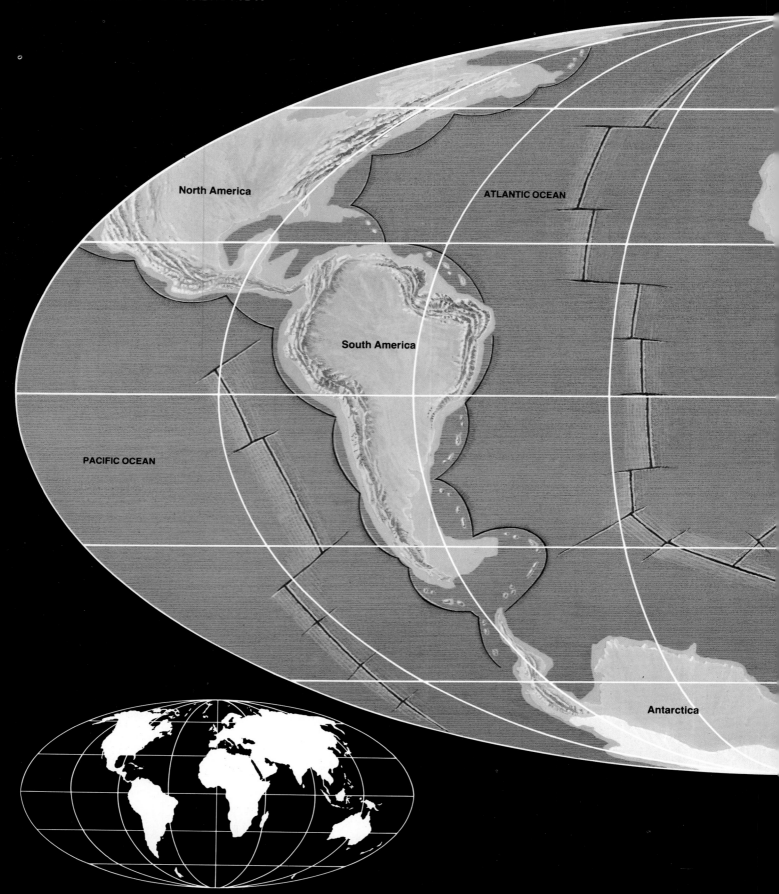

North America

ATLANTIC OCEAN

South America

PACIFIC OCEAN

Antarctica

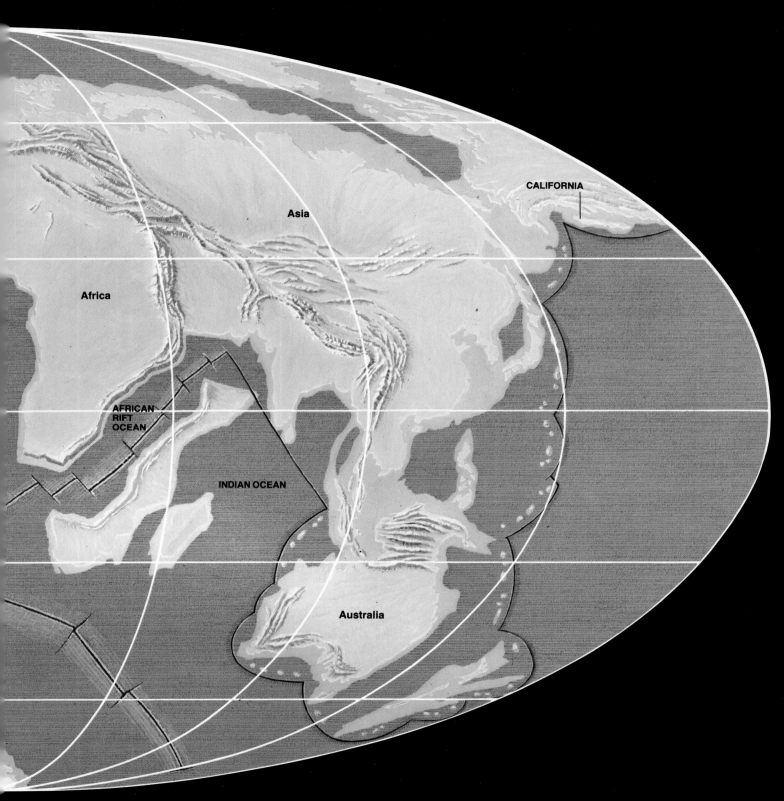

If present plate movements continue for 100 million years, the Atlantic will become the earth's largest ocean, far surpassing the shrunken Pacific. Africa will grind into Europe, closing the Mediterranean and building new mountains, while continued rifting will turn Africa's Great Rift Valley into a seaway. Other crustal movements will shave a sliver of California from the North American coast and drag Australia toward Southeast Asia, folding the intervening islands into collision mountains. New plate movements will begin, and subduction trenches may develop along the Atlantic coasts of the Americas and around Australia, as old, dense sea floor, laden with sediments, sinks into the mantle. Volcanic island arcs and tall coastal mountains will appear on the landward sides of the trenches.

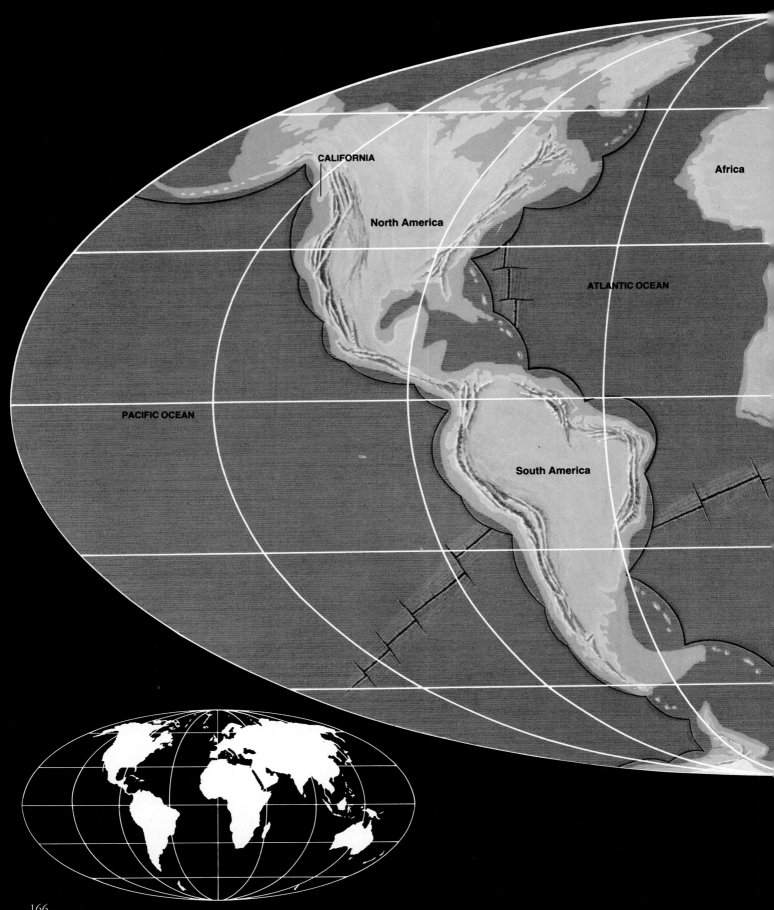

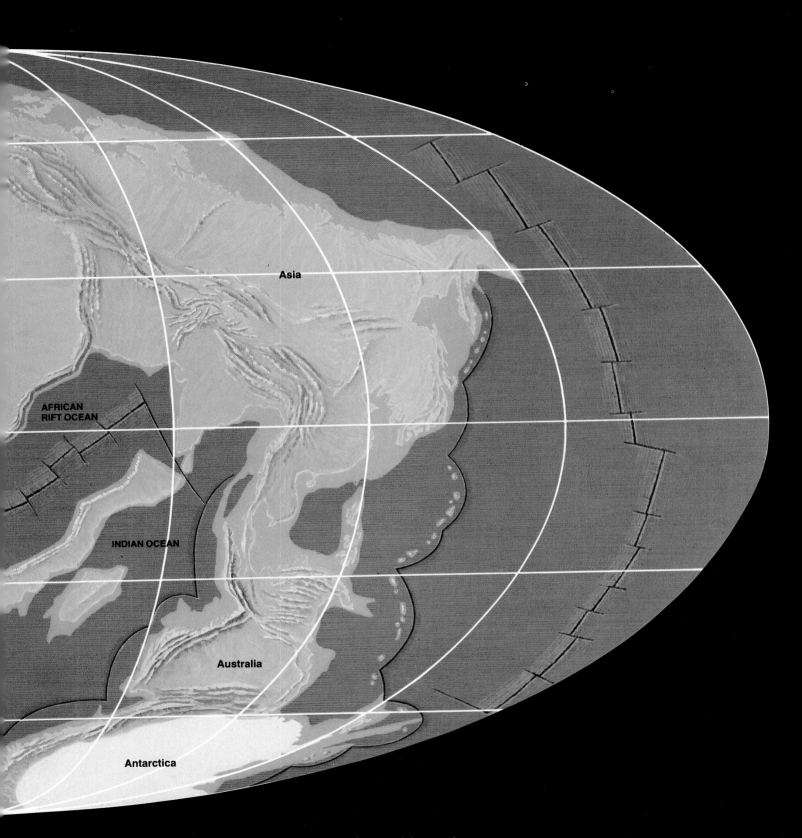

Asia

AFRICAN
RIFT OCEAN

INDIAN OCEAN

Australia

Antarctica

As new subduction zones consume oceanic crust, the Atlantic and Indian Oceans will narrow. New rifts may expand the Pacific once again, while the rift ocean sundering the African continent continues to grow. As North America and Asia separate, a scrap of Siberia will ride with Alaska. California's journey may end as the subduction of the plate on which it rides crumples it against the coast of Alaska and incorporates it into new mountain chains. Another subduction zone, south of Australia, will draw Antarctica north, welding it to Australia and folding up mountains along the continental splice.

250 Million Years from Now

Africa
North America
South America

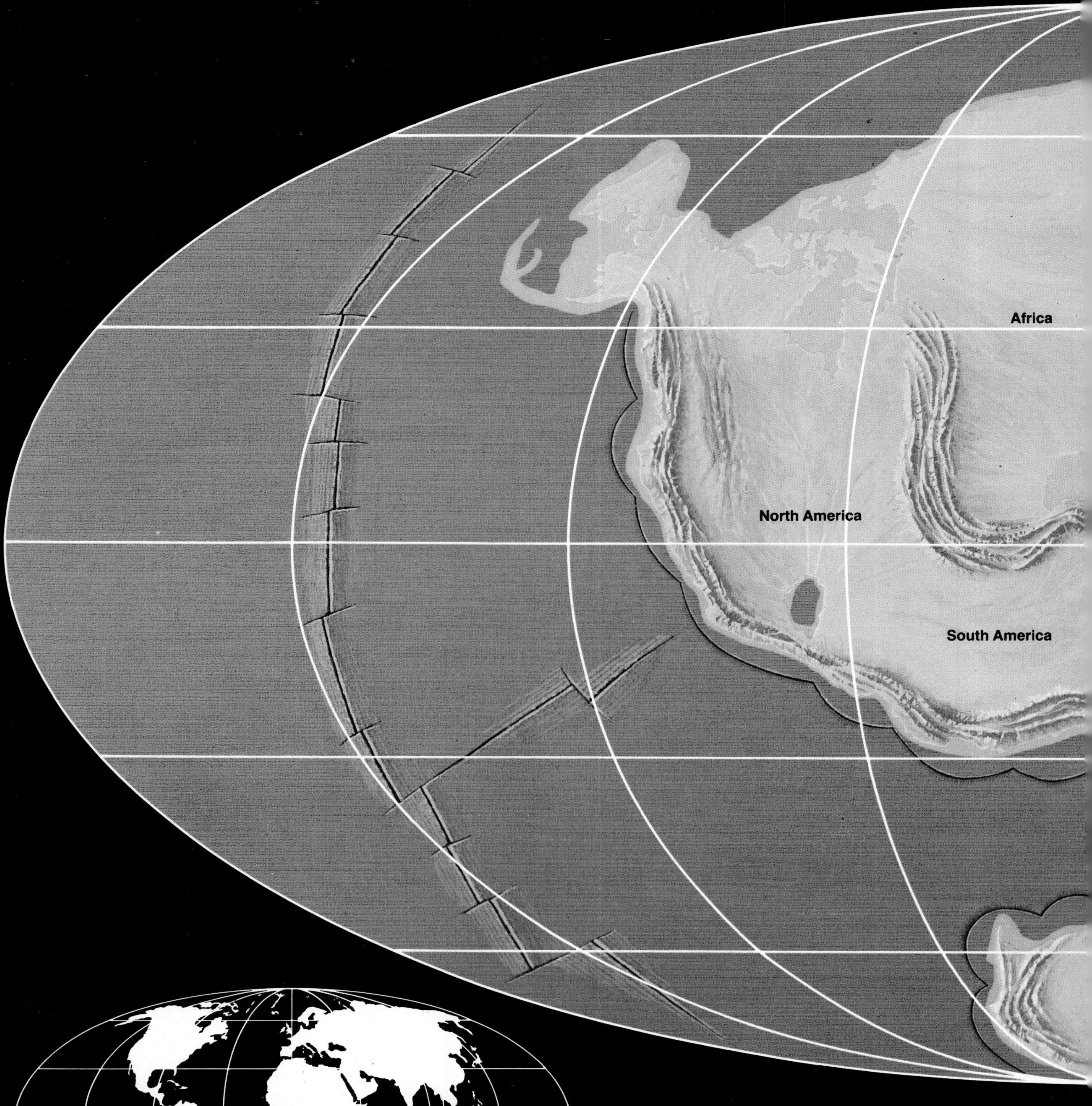

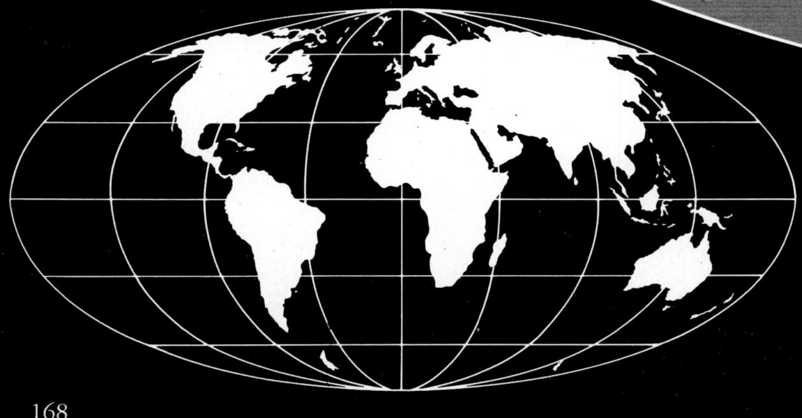

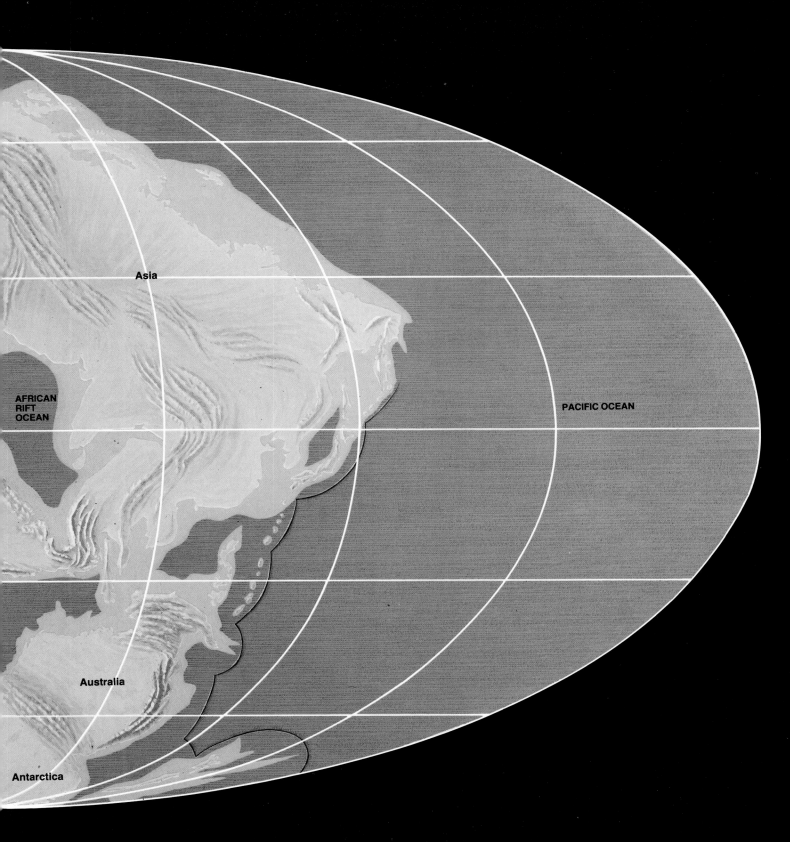

Asia

AFRICAN
RIFT
OCEAN

PACIFIC OCEAN

Australia

Antarctica

Continued subduction of the ocean floors may
close the Atlantic and Indian Oceans,
drawing the continents together in a single
immense land mass sprawling across half
the globe—a new Pangaea. Ranges of folded
mountains heaved up by the continental
collisions will mark former coastlines. The
Pacific Ocean will once again be dominant, still
spreading outward from central rifts and
descending into subduction trenches that ring
the supercontinent. Part of the now-ancient
African rift ocean will remain trapped within the
huge land mass—an ocean penned in by plate
movements, much like the Mediterranean today.

ACKNOWLEDGMENTS

For their help in the preparation of this book the editors wish to thank: **In Denmark:** Copenhagen—Dr. Niels Hald, Director, Geologisk Museum ved Københavns Universitet. **In France:** Grenoble—Jérôme Biju-Duval; Marcel Lemoine and George Mascle, Institut Dolomieu, Université de Grenoble; Orléans—Jacques Varet, B.R.G.M.; Orsay—Robert Brousse, Université d'Orsay; Paris—Jacques Angelier, Jacques Bourgois, Vincent Courtillot and Xavier Le Pichon, U.E.R. des Sciences de la Terre, Université-Paris VI; Jean-Louis Cheminée and Paul Tapponnier, Institut de Physique du Globe, Université-Paris VI; Jean Pierre Peulvast, Université Paris-Sorbonne; Haroun Tazieff; Rennes—Pierre Choukroun, Université de Rennes; Rueil-Malmaison—Bernard Biju-Duval, Institut Français du Pétrole. **In Great Britain:** Cambridge—Dr. Geoffrey King, Bullard Laboratories, University of Cambridge; Edinburgh—Professor G. Y. Craig, Grant Institute of Geology, University of Edinburgh; Hove, Sussex—Mrs. Doris L. Holmes; London—Professor Ronald Mason, Royal School of Mines, University of London; Sheffield—Professor Keith Miller, University of Sheffield. **In Italy:** Naples—Alessandro Oliveri del Castillo, Istituto di Geologia e Geofisica, Università degli Studi di Napoli. **In Japan:** Tokyo—Yoichi Ohta, Japan Petroleum Exploration Co., Ltd.; Dr. Kazuaki Nakamura, Earthquake Research Institute, and Dr. Hitoshi Takeuchi, Professor Emeritus, University of Tokyo. **In the Netherlands:** Amsterdam—Dr. Derk Jongsma, Dr. Jan E. van Hinten; The Hague—Dr. Rein W. van Bemmelen; Leyden—Dr. Henk Schalke; Ultrecht—Dr. Nicolaas J. Vlaar. **In South Africa:** Cape Town—Dr. A. R. Newton, University of Cape Town. **In Sweden:** Uppsala—Professor Hans Ramberg, Institute of Geology, University of Uppsala. **In Switzerland:** Neuchâtel—Francis Persoz, Institut de Géologie, Université de Neuchâtel. **In the United States:** California—(Berkeley) Walter Alvarez, University of California at Berkeley; (La Jolla) Deborah Day, Thomas H. Jordan, Jacqueline Parker, Dr. Melvin Peterson, William Smithey, Scripps Institution of Oceanography; (Menlo Park) David L. Jones, U.S. Geological Survey; (Pasadena) Kenneth D. Graham, Dr. Michael Kobrick, California Institute of Technology; (Riverside) Michael O. Woodburne, University of California; (San Francisco) Andrew Kramer, The Burdick Group; Connecticut—(New Haven) Judith Ann Schiff, Yale University Library; District of Columbia—Dr. Jon D. Dykstra, Earth Satellite Corporation; Dr. Edward A. Flinn, NASA; Barbara A. Shattuck, National Geographic Society; Illinois—(Champaign/Urbana) Albert V. Carozzi, University of Illinois at Champaign/Urbana; (Chicago) Christopher Scotese, University of Chicago; James Vaughan, The John Crerar Library; Louisiana—(Lafayette) Ross A. Brupbacher, Domengeaux & Wright; Maryland—(Annapolis) Greg Harlin, Rob Wood, Stansbury, Ronsaville, Wood; (Baltimore) Bruce D. Marsh, The Johns Hopkins University; (Bethesda) Fred Bigio, B-C Graphics; (Bowie) I'Ann Blanchette; (Gaithersburg) Jaime Quintero, Carol Schwartz; (Greenbelt) Paul D. Lowman Jr.; Massachusetts—(Cambridge) Dr. Ursula Marvin, Smithsonian Astrophysical Observatory; Dr. John G. Sclater, Massachusetts Institute of Technology; Dr. Robert D. Ballard, John Donnelly, Nancy Green, Dr. J. R. Heirtzler, Dr. John A. Whitehead Jr., Woods Hole Oceanographic Institution; New Jersey—(Princeton) Dr. Hollis Hedberg; New York—(Albany) Dr. Kevin Burke, State University of New York; (New York) Graham R. Kimber, Exxon Corporation; (Palisades) Annie Lewis, John Mutter, Walter Pitman, Marie Tharp, Lamont-Doherty Geological Observatory; Ohio—(Columbus) Dr. William Zinsmeister, Ohio State University; Pennsylvania—(Maytown) Ken Townsend; (State College) Dr. Robert Scholten, Pennsylvania State University; Rhode Island—(Narragansett) Dr. Paul Jeffrey Fox, University of Rhode Island; Virginia—(Burke) Paul Salmon; (Reston) Harold L. Burstyn, William D. Carter, Dr. John Filson, Walter W. Hayes, Clifford M. Nelson, Mary C. Rabbitt, Dr. Henry Spall, Richard S. Williams Jr., Isidore Zietz, U.S. Geological Survey; Washington—(Winthrop) John C. Holden, International Stop Continental Drift Society; Wyoming—(Laramie) Dr. Kip Vernon Hodes, University of Wyoming. **In West Germany:** Bensberg Bei Köln—Dr. Martin Schwarzbach; Bonn—Alexander von Humboldt Foundation; Dr. Hanno Beck; Clausthal-Zellerfeld—Dr. Andreas Pilger and Dr. Artur Rösler, Geologisches Institut; Dr. Peter Halbach, Mineralogisch-Petrographisches Institut; Eichstätt—Dr. Herman Holzbauer, Director, Universitätsbibliothek; Hamburg—Dieter Steiner, *GEO;* Leipzig—Dr. Heinz Lüdemann, Director, Institute of Geography and Geoecology, Academy of Sciences of the GDR; Munich—Charlotte Wegener; Waldsee/Pfalz—Franz-Karl Freiherr von Linden; West Berlin—Dr. Roland Klemig and Heidi Klein, Bildarchiv Preussischer Kulturbesitz.

The editors also wish to thank the following persons: Wibo van der Linde, Amsterdam; Robert Gilmore, Auckland; Lois Lorimer, Copenhagen; Peter Hawthorne, Johannesburg; Trini Bandres, Madrid; John Dunn, Melbourne; Cornelius Verwaal, New York; Mary Johnson, Stockholm; Traudl Lessing, Vienna; Marie Gerald, Wellesley Hills.

Particularly useful sources of information and quotations used in this volume were: *The Earth in Decay: A History of British Geomorphology, 1578-1878* by Gordon L. Davies, American Elsevier Publishing Company, 1969; *A Revolution in the Earth Sciences: From Continental Drift to Plate Tectonics* by A. Hallam, Oxford, Clarendon Press, 1973; *Continental Drift: The Evolution of a Concept* by Ursula B. Marvin, Smithsonian Institution Press, 1973; *The Origins of Continents and Oceans* by Alfred Wegener, Dover Publications, 1966; and *The Floor of the Sea: Maurice Ewing and the Search to Understand the Earth* by William Wertenbaker, Little, Brown, 1974.

The index was prepared by Gisela S. Knight.

BIBLIOGRAPHY

Books

Argand, Emile, *Tectonics of Asia.* Ed. and transl. by Albert V. Carozzi. Hafner Press, 1977.

Atlas of the Oceans. Rand McNally, 1972.

Attenborough, David, *Life on Earth: A Natural History.* London: Collins, British Broadcasting Corp., 1979.

Bárðarson, Hjálmar R., *Ice and Fire.* Reykjavik: Hjálmar R. Bárðarson, 1971.

Beiser, Arthur, and the Editors of Time-Life Books, *The Earth.* Time-Life Books, 1962.

Bird, John M., ed., *Plate Tectonics: Selected Papers from Publications of the American Geophysical Union.* American Geophysical Union, 1980.

Booth, Basil, and Frank Fitch, *Earthshock.* London: J. M. Dent & Sons Ltd., 1979.

Botting, Douglas, *Humboldt and the Cosmos.* Harper & Row, 1973.

Briggs, Peter, *200,000,000 Years beneath the Sea.* Holt, Rinehart and Winston, 1971.

Burnet, Thomas, *The Sacred Theory of the Earth.* London: 1684.

Calder, Nigel, *The Restless Earth: A Report on the New Geology.* Viking Press, 1972.

Cohen, I. Bernard, *Album of Science: From Leonardo to Lavoisier, 1450-1800.* Charles Scribner's Sons, 1980.

Colbert, Edwin H., *Wandering Lands and Animals.* E. P. Dutton, 1973.

Cox, Allan, *Plate Tectonics and Geomagnetic Reversals.* W. H. Freeman and Company, 1973.

Davies, Gordon L., *The Earth in Decay: A History of British Geomorphology, 1578-1878.* American Elsevier Publishing Company, 1969.

Dennis, John G., *Structural Geology.* John Wiley & Sons, 1972.

Dictionary of Scientific Biography. Ed. by Charles Coulston Gillispie, et al. Charles Scribner's Sons, 1981.

Dineley, David, *Earth's Voyage through Time.* Alfred A. Knopf, 1974.

Du Toit, Alex. L., *Our Wandering Continents: An Hypothesis of Continental Drifting.* Greenwood Press, 1974.

Earle, Sylvia A., and Al Giddings, *Exploring the Deep Frontier: The Adventure of Man in the Sea.* National Geographic Society, 1980.

Fridriksson, Sturla, *Surtsey: Evolution of Life on a Volcanic Island.* London: Butterworths, 1975.

Gass, I. G., Peter J. Smith, and R.C.L. Wilson, eds., *Understanding the Earth: A Reader in the Earth Sciences.* M.I.T. Press, 1972.

Glen, William, *Continental Drift and Plate Tectonics.* Charles E. Merrill, 1975.

Golden, Frederic, *The Moving Continents.* Charles Scribner's Sons, 1972.

Gross, M. Grant, *Oceanography: A View of the*

Earth. Prentice-Hall, 1977.

Hallam, A., *A Revolution in the Earth Sciences: From Continental Drift to Plate Tectonics.* Oxford: Clarendon Press, 1973.

Heezen, Bruce C., and Charles D. Hollister, *The Face of the Deep.* Oxford University Press, 1971.

Heintze, Carl, *The Bottom of the Sea and Beyond.* Thomas Nelson, 1975.

Idyll, C. P., ed., *Exploring the Ocean World: A History of Oceanography.* Thomas Y. Crowell, 1972.

Jacobs, J. A., R. D. Russell, and J. Tuzo Wilson, *Physics and Geology.* McGraw-Hill, 1974.

Kay, John, *A Series of Original Portraits and Caricature Etchings,* Vol. 1. Edinburgh: Adam and Charles Black, 1877.

Knox, R. Buick, *James Ussher—Archbishop of Armagh.* Cardiff: University of Wales Press, 1967.

Lambert, David, *Dinosaurs.* Crown Publishers, 1978.

Levin, Harold L., *The Earth through Time.* W. B. Saunders Company, 1978.

Limburg, Peter R., *Oceanographic Institutions: Science Studies the Sea.* Elsevier/Nelson Books, 1979.

McPhee, John, *Basin and Range.* Farrar, Straus, Giroux, 1981.

Magnússon, Magnús, *Iceland.* Reykjavik: Almenna Bókafélagið, 1978.

Marvin, Ursula B., *Continental Drift: The Evolution of a Concept.* Smithsonian Institution Press, 1973.

Miller, Keith, *Continents in Collision.* George Philip, 1982.

Motz, Lloyd, ed., *The Rediscovery of the Earth.* Van Nostrand Reinhold, 1982.

Nelken, Halina, *Alexander von Humboldt: His Portraits and Their Artists.* Berlin: Dietrich Reimer Verlag, 1980.

Phinney, Robert A., *The History of the Earth's Crust.* Princeton University Press, 1966.

Powell, John Wesley, *Exploration of the Colorado River of the West and Its Tributaries.* U.S. Government Printing Office, 1875.

Rabbitt, Mary C., *Minerals, Lands, and Geology for the Common Defence and General Welfare,* Vol. 1, *Before 1879.* U.S. Government Printing Office, 1979.

Rudwick, Martin J. S., *The Meaning of Fossils: Episodes in the History of Palaeontology.* American Elsevier, 1972.

Runcorn, S. K., ed., *Continental Drift.* Academic Press, 1962.

Scientific American:
Continents Adrift and Continents Aground. W. H. Freeman and Company, 1976.
The Ocean, W. H. Freeman and Company, 1969.
Oceanography, W. H. Freeman and Company, 1971.

Shor, Elizabeth Noble, *Scripps Institution of Oceanography: Probing the Oceans, 1936-1976.* Tofua Press, 1978.

Smyth, Albert Henry, *The Writings of Benjamin Franklin,* Vol. 8, *1780-1782.* Macmillan, 1906.

Sullivan, Walter, *Continents in Motion: The New Earth Debate.* McGraw-Hill, 1974.

Tarling, Don and Maureen, *Continental Drift: A Study of the Earth's Moving Surface.* Anchor Press/Doubleday, 1975.

Thorndike, Joseph J., Jr., ed., *Mysteries of the Deep.* American Heritage, 1980.

Uyeda, Seiya, *The New View of the Earth: Moving Continents and Moving Oceans.* Transl. by Masako Ohnuki. W. H. Freeman and Company, 1971.

Von Linden, Franz-Karl, and Helfried Weyer, *Iceland.* Berne: Kümmerly & Frey, 1974.

Wegener, Alfred, *The Origin of Continents and Oceans.* Transl. by John Biram. Dover Publications, 1966.

Wegener, Else, ed., *Greenland Journey: The Story of Wegener's German Expedition to Greenland in 1930-31 as Told by Members of the Expedition and the Leader's Diary.* Transl. by Winifred M. Deans. Glasgow: Blackie & Son, 1939.

Wendt, Herbert, *Before the Deluge.* Transl. by Richard and Clara Winston. Doubleday, 1968.

Wertenbaker, William, *The Floor of the Sea: Maurice Ewing and the Search to Understand the Earth.* Little, Brown, 1974.

Young, Patrick, *Drifting Continents, Shifting Seas: An Introduction to Plate Tectonics.* Franklin Watts, 1976.

Periodicals

Alexander, Tom:
"A Revolution Called Plate Tectonics Has Given a Whole New Understanding of the Configuration of the Earth—and How and Why It Came About." *Smithsonian,* January 1975. First of two articles.
"Plate Tectonics Has a Lot More to Tell Us about the Earth as It Is—and as It Will Be." *Smithsonian,* February 1975. Second of two articles.
"The Secret of the Spreading Ocean Floors." *Fortune,* February 1969.

Alvarez, Luis W., Walter Alvarez, Frank Asaro and Helen V. Michel, "Extraterrestrial Cause for the Cretaceous-Tertiary Extinction." *Science,* June 6, 1980.

Ballard, Robert D.:
"Notes on a Major Oceanographic Find." *Oceanus,* Summer 1977.
"Window on Earth's Interior." *National Geographic,* August 1976.

Ballard, Robert D., and J. Frederick Grassle, "Return to Oases of the Deep." *National Geographic,* November 1979.

"Challenger Drills into Basement." *Geotimes,* May 1982.

Ewing, Maurice:
"Exploring the Mid-Atlantic Ridge." *National Geographic,* September 1948.
"New Discoveries on the Mid-Atlantic Ridge." *National Geographic,* November 1949.

George, Uwe, "Africa: An Ocean Is Born." *GEO,* Vol. 1, premier issue.

Golden, Frederic, "Strange Creatures of the Deep." *Science,* May 24, 1982.

Henbest, Nigel, "127 Minutes under Venus's Orange Skies." *New Scientist,* March 11, 1982.

Herrman, Lisa, "Alien Worlds on the Ocean Floor." *Science Digest,* April 1981.

Hessler, Dr. Robert, "Oasis under the Sea—Where Sulphur Is the Staff of Life." *New Scientist,* December 10, 1981.

Hilts, Philip J., "Antarctica Bones Give Evidence of Continental Drift." *The Washington Post,* March 21, 1982.

Hsü, Kenneth J.:
"K-T Event Debated; Conclusion:?!" *Geotimes,* July 1982.

"When the Mediterranean Dried Up." *Scientific American,* December 1972.

Hurley, Patrick M., "The Confirmation of Continental Drift." *Scientific American,* April 1968.

Kerr, Richard A., "The Bits and Pieces of Plate Tectonics." *Science,* March 7, 1980.

Kertz, Walter, Karl Weiken, Klaus Strobach, and Hans Dieter Heck, "Alfred Wegener hatte recht: Die Kontinente bewegen sich doch." *Bild der Wissenschaft,* November 1980.

Kobrick, Michael, "Topography of the Terrestrial Planets." *Astronomy,* May 1982.

Mantura, Andrew J., "The Mediterranean Enigma." *Oceans,* January 1977.

Matthews, Samuel W., "This Changing Earth." *National Geographic,* January 1973.

"Mediterranean: Probing Its Origins." *Science News,* July 12, 1975.

Merritt, J. I., "Hess's Geological Revolution." *Princeton Alumni Weekly,* September 24, 1979.

Raff, Arthur D., "The Magnetism of the Ocean Floor." *Scientific American,* October 1971.

Russell, Dale A., "The Mass Extinctions of the Late Mesozoic." *Scientific American,* January 1982.

Schmincke, Hans-Ulrich, "Detectives of the Deep." *GEO,* June 1981.

Schraps, Wolfgang, "Iceland." *GEO,* March 1981.

"The Seams of the Earth." *Mosaic,* November-December 1975.

Simmons, Henry, "Before Pangea." *Mosaic,* March-April 1981.

"Strange World without Sun." *National Geographic,* November 1979.

Sullivan, Walter, "Drill Hole Yields Clue to Structure of Ocean Bed." *The New York Times,* July 24, 1979.

"Tectonics on Venus: Like That of Ancient Earth?" *Science,* January 15, 1982.

Teske, Richard, "Asteroid Collisions with Earth." *Astronomy,* January 1982.

Thomsen, D. E., "The Other Side of Venera: Capturing Venusian Evolution?" *Science News,* April 10, 1982.

Thorarinsson, Sigurdur, "Surtsey: Island Born of Fire." *National Geographic,* May 1965.

West, Susan:
"Alaska: The Fragmented Frontier." *Science News,* January 3, 1981.
"A Patchwork Earth." *Science 82,* June 1982.
"Smokers, Red Worms and Deep Sea Plumbing." *Science News,* January 12, 1980.

"What Drives the Earth's Plates?" *Mosaic,* September/October 1979.

Other Publications

"Alfred Wegener, 1880-1930, Leben und Werk." Exhibition on the occasion of the 100th anniversary of his birthday. 1980.

Dietz, Robert S., "Plate Tectonics: Past, Present and Future," *1978 Yearbook of Science and the Future.* Encyclopedia Britannica.

Holmes, Arthur, "Radioactivity and Earth Movements." Transactions of the Geological Society of Glasgow, 1928-1929.

Humboldt, Alexander von, "Esquisse d'Un Tableau Géologique de l'Amérique Méridionale," *Journal de Physique, de Chimie, d'Histoire Naturelle,* Vol. 53, 1801.

"Plate Tectonics and Man." Reprint from the U.S. Department of the Interior/Geological Survey Annual Report, 1976.

INDEX

Mammoth Hot Springs, Yellowstone National Park, *142-143*

Mantle, earth, 19, 43, 80, 101, 133-134; basalt layer, 25; convection in, 134-135, *diagram* 137, 139; convection cell models, 135, *diagram* 138; deep-sea drilling project, 80, *81;* hot spots, 136-137; plate subduction into, 108, 115, 117, 122, 124, 143; upper, 124; upwelling magma, 108, 117. *See also* Magma

Marianas Trench, 110

Marine life: in deep-sea rift, *119-121;* fossil finds in mountains, 7, 10, *12-13,* 14, 15, 17, 153; Pangaean, 145, 154; post-Pangaean, 155

Marsupials, 156-*157, 158-159*

Mason, Ronald G., 84, 87

Massachusetts Institute of Technology, 106, 143

Matthews, Drummond, 85-86, 87

Maxwell, Arthur E., 91

Mediterranean Sea, 110, 131, 148, 153, 154, *map* 163; future of, *map* 164-165

Meinesz, Felix A. Vening, 80

Menard, Henry, 90

Mesosaurus fossil finds, 19, *map* 156

Microplates, 123, 131

Mid-Atlantic Ridge, 29, 69, 72, 110, 126; birth of, 153; exploration and mapping of, 74-75, 77, 91, 92, *98-99,* 117; length of, 61; magnetic profile at, 88; mineral formation at, 159; parallelism to continental coasts, 76, *map* 100; sea-floor spreading at, 88, 102-103, 110; surface manifestations of, 61; volcanism, 61, 74, 92

Mid-Atlantic rift, 29, *68-69,* 74, 77, 110; descent by submersibles into, 117; earthquake zone, 74-75; heat flow, 75, 77; magma flow, 61, 63; width of, 75

Mid-Ocean Ridge (system), 7, 75-77, *map* 78-79, 80, 85, 86, 102, 108, 124, *diagram* 126-127, 139; earthquake zones, 75, 89, 103; forming of new crust at, 108, 110; heat at, 75, 77, 108, 139; height of, 126; hot spots beneath, 136-137, 139; interdependent system, 103; length of, 75, 79, 126; magnetic profiles at, 88-90, 102; mineral formation at, 159; offset fractures, 88-89, *diagram* 90, 103, 110; origin of, 148; transform faults, 89, *diagram* 90, 103, 110; undersea life-forms, 119; width of, 126. *See also* Sea-floor spreading

Midocean rifts, 71, 75-76, 110, *diagram* 124-127, 139

Mineral ores: mountain deposits, 143, 161; on sea floor, 149, *152-153,* 159

Mohole, Project, 80, *81*

Mohorovičić, Andrija, 80

Moon, theories of origin of, 14, 28, 29

Moran, Thomas, painting by, *26-27*

Morgan, W. Jason, 102-103, 137

Morley, Lawrence W., 86

Mountains and mountain chains, 7, 29, *diagram* 124-125, *132, 160;* African-South American match, 16, 46; ages of, 43; birth of, Pangaea, 145, *map* 162; birth of, post-Pangaea, 153, *map* 163; buoyancy of, 25, 28, 115; contracting-earth theory as explanation of, 18-19, 43, 48; European-North American match, 16, 43, 53, 145; folding of, *22-23,* 48, *diagram* 53, 54-55, 103, 110, 115, *diagram* 124-125, *diagrams* 128-129; folding model, *21;* layered, former ocean crust in, *146-147;* marine fossils found in, 7, 10, *12-13,* 14, 15, 17, 153; microplate

accumulations in, 123, 131; mineral ore deposits in, 143, 161; ophiolite zones, 143; plate borders obscured by, 131

N

National Geographic Society, 72

National Science Foundation, 88

Natron, Lake, *34-35*

Nature (magazine), 28, 86

Nazca Plate, 107, 135; movement of, 115, 131

Neumayr, Melchior, 20

New Hebrides, 115

New Zealand, 153

Niigata, Japan, *106*

North America, 123; drift of, 48-49, 83; former locations and movements of, 145, 148, 153, *maps* 162-163; fossil match with Europe, 14, 41; geologic formation match with Europe, 48; joining to South America, 154, 157-158; mammals, 156, 157-158; mountain chain match with Europe, 16, 43, 53, 145; once joined to Africa, 143, 145; Pacific crust pieces found on land, 122-123, *146-147;* as part of Laurasia, 56, 145, 148, 156, *map* 162; possible future of, *maps* 164-167; separation from Eurasia, 153, 157; in supercontinent theories, 20, 56, 83, 145, *map* 162; trans-Bering land bridge, 157

North American Plate, 7, 69, 105, 107, 131; movement of, 110, 139

North Atlantic Ocean, origin of, 148, 153, *map* 162

Norway, 53; sandstone band, 48

Novum Organum (Bacon), 10

Nur, Amos, 123

O

Ocean circulation, and weather, 145

Ocean floor, 72, 80, 124; age and depth relationship, 122; age of rocks of, 74, 77, 80; deepest point, 110; magnetic profile of, 84-85, 87-90; mapping of, 72, 74, 75; mineral precipitation on, 149, *152-153;* sedimentation of, *72, 73-74,* 77, 80, 91; seismic profile, *84-85. See also* Sea-floor spreading; Subduction

Oceanic crust, 80, 107-108; composition, 43, 74, 75; forming of new, 108, 110, 117, 124; high density and weight, 25, 28, 43, 46, 74, 110; pieces found on land, 122-123, 128, 143, *146-147;* thickness of, 74, 77, 80

Oceanic plates, 107, 124; collisions of, 110; collision with continental plates, 110, 115, *diagram* 128, 161; plate border identification, 131; rate of movement, 131; subduction of, 110, 115, 117, *diagram* 124-125, 128, *diagram* 129; thermal contraction, 122

Oceans, 7; birth of, 103, 108, 124, *diagrams* 126-127, 143, 148; closing up of, 110, *diagram* 128, 129, 143; depth of, 122; early theories of origin of, 11, *14,* 16-17, 18-19, 28, 43; fixed, theory of, 18, 29, 80, 86; "geopoetry" view (Hess), 77, 80, 82, 86, 90; universal (Panthalassa), 143

Ocean trenches, 80, *84-85,* 103, 115, *diagram* 124-125; dimensions, 110; earthquake and volcanic activity along, 103, 115, 128; flip of, 115; forming of, 110, *diagram* 128; location near continents, 115; sinking of plates into, 115, *diagram* 128, 139. *See also* Subduction

O'Connell, Richard, 133, 135

Offshore oil, 72, 159

Oliver, Jack, 103, 106

"On Some of the Greater Problems of Physical

Geology" (Dutton), 28

Ophiolites, ophiolite zones, 143, 146-147

Origin of Continents and Oceans, The (Wegener), 46, 49, *50,* 52

Our Wandering Continents (Du Toit), 56

Owen, H.M.S., 85

P

Pacifica, 123

Pacific Ocean: cobalt mining, *149-153;* crust pieces found on North American land, 122-123, *146-147;* deep-sea geysers, *118-119;* deep-sea life-forms, *119-121;* earthquake zones, 89, 102; guyots, 76-77; hot-spot island chains, 136-137; island arcs, 115, *116-117;* magnetic profiles, 84-85, 87, 88-90; Mid-Ocean Ridge, 76, 159; mineral formation, 159; narrowing of, 108; origin of, 103; possible future of, *maps* 164-169; relationship of floor's age to depth, 122; Ring of Fire, 115, 140; sea-floor spreading, 71, 88; theories on origin of, 28; volcanism, 115, 136-137, 140, 154. *See also* East Pacific Rise

Pacific Plate, 7, 105, 107, 110; hot spots beneath, 136-137; length of, 139; movement of, 110, 131, 136; seismic profile, *84-85;* thickness of, 139

Pakistan: Karakoram Range, *132;* Sulaiman Range, *160*

Paleomagnetism, 82-86, 123, 143, 162

Palisse, Bernard, 10

Panama, Isthmus of, 154, 157

Pangaea, 48, *map* 50, 143, 145; birth of, 143-145, *map* 162; breakup of, 148, 155, 159, *map* 162, 163; fossils found in Himalayas, 153; life-forms of, 145, 148, 154, 155; meaning of term, 162; predecessors of, 143, *map* 162

Panthalassa, 143, 148

Parker, Robert L., 102, 103

Peru-Chile Trench, 15

Philippines, 115; Mount Taal, *116*

Physics of the Earth's Crust (Fisher), 134

Pillow lava, *70, 74, 98,* 117, 122

Pioneer Venus space probe, 144, 145

Pitcairn Island, 136

Placet, François, 10

Plate rifting, 110, *diagram* 124-125, 126, *diagram* 127, 141, 148; triple rift, *diagram* 108, *109*

Plate tectonics, 7, 107-117, 122-123, *diagrams* 124-131; beginnings of, 143; development and confirmation of theory, 102-107; driving-force hypotheses, 108, 133-139, 143; geometrical evidence, 102-103; origin and meaning of term, 106; seismological evidence, 103. *See also* Tectonic plates; Tectonic processes

Playfair, John, 15

Poles, magnetic, 82, 83-84; reversal, 84, 86, 88

Powell, John Wesley, 21, *24,* 25, 27

Pozzuoli (Italy), Temple of Jupiter, *46-47*

Press, Frank, 139

Princeton University, 76, 102

Principles of Geology (Lyell), 18

"Proof of Ocean-Floor Spreading?" (Vine), 90

Protocontinents, 24, 29, 162; Atlantis and Gondwanaland (Suess), 20; Gondwanaland and Laurasia (Du Toit), 56; Wegener's, 48, 143. *See also* Supercontinents

R

Radioactive dating, 43, 84, 106

Radioactive decay, heat of, 21, 28, 75, 108, 115

Printed in Spain by Novograph, S. A. Madrid
Depósito Legal: M-37290-XXX